GW01230013

The Theory of Peasant Co-operatives

The Theory of Peasant Co-operatives

ALEXANDER CHAYANOV

Translated by
David Wedgwood Benn

Introduction by
Viktor Danilov

I.B. Tauris & Co Ltd
Publishers
London · New York

Published in 1991 by
I.B.Tauris & Co Ltd
110 Gloucester Avenue
London NW1 8JA

First published in Russian under the title *Osnovnye idei i formy organizatsii sel'skokhozyaistvennoi kooperatsii* (*The basic ideas and organizational forms of agricultural co-operation*), Moscow 1927. This was the second edition, revised and supplemented, of a book first published in 1919.

NOTE: Some of the tables in Alexander Chayanov's original Russian text contain arithmetic inaccuracies. Since it has not been possible to check Chayanov's figures from the 1920s, the tables have been left as the author prepared them. In the English edition of the book certain cuts have been made and some tables and diagrams have been omitted.

Copyright to English translation © 1991 I.B.Tauris

All rights reserved. Except for brief quotations in a review, this book, or any part thereof, must not be reproduced in any form without permission in writing from the publisher.

A CIP record for this book is available from the British Library.

ISBN 1-85043-189-2

Printed and bound in Great Britain by
WBC Print Ltd, Bridgend, Mid Glamorgan

Contents

Foreword to the Second World Series by Teodor Shanin		vii
Introduction by Viktor Danilov		xi
Author's Comment		xxxvii
1	The Processes and the Concept of Vertical Concentration in the Rural Economy: Peasant Co-operation as an Alternative	1
2	The Theory of Differential Optima and Co-operatives in the Peasant Economy	24
3	Credit in the Peasant Economy	53
4	Co-operative Credit Societies	72
5	The Peasant Family's Money Economy and its Organization on Co-operative Principles	91
6	The Basic Principles of the Co-operative Organization of Commodity Circulation	115
7	The Organization of Co-operative Marketing and Reprocessing Enterprises	131
8	Machinery Users' Associations	147
9	Dairy Farming Reprocessing and Cattle-rearing Co-operatives	162
10	Peasant Co-operation for the Purpose of Cattle-rearing	183

11	Co-operative Insurance	187
12	Associations Concerned with Land	196
13	Collective Farms or 'Total Agricultural Co-operation'	207
14	The Basic Principles of Organization of Agricultural Co-operatives	224

The Second World Series

'In the West they simply do not know Russia . . . Russia in its germination.'

Alexander Hertzen

As a publication project *The Second World* pursues an explicit goal, admits to a bias and proceeds on a number of assumptions. This should be stated at the outset. The series will aim to let the Soviet authors and their historical predecessors in tsarist Russia speak with their own voices about issues of major significance to us and to them. It will focus particularly on their explorations of their own society and culture, past and present, but set no rigid boundaries to these; some of the texts will be more general while others will carry primary evidence, for example, memoirs, documents, etc. Many of the texts have been commissioned to reflect the most recent issues and controversies of Gorbachev's *perestroika*.

To bridge differences of culture and experience each of the books will carry a substantial introduction by a Western scholar within the field. Particular care will also be taken to maintain satisfactory standards of translation and editing.

A word about words. A generation ago the term 'Third World' was coined in its current meaning, to indicate a somewhat imprecise combination of societal characteristics – the post-colonial experience, under-industrialization, relative poverty and the effort to establish an identity separate from the superpowers, the 'Bandung camp'. This left implicit yet clear which were the other two 'worlds'. It was 'us' and 'them', those best represented by the USA and those best represented by the USSR. Much has changed since, giving the lie to crude categorizations. But in research and the media, at the UN and on television, the words and the meanings established in the 1960s are still very much with us. This makes the title of our project intelligible to all, yet, hopefully, should also make the reader pause for a moment of reflection.

Turning to the invisible rules and boundaries behind the editorial selection let us stress first the assumption of considerable social continuity between pre-revolutionary and post-revolutionary societies. Present without past is absurd (as is, of course, the treatment of the USSR as simply the Russia of old). Next, to talk of pre-revolutionary Russia/USSR is not simply to talk of the Russians. The country is multi-ethnic, as have been its intellectual achievements and self-evaluations. Yet all the books presented are deeply embedded in Russian language and cultural traditions. Lastly, we shall aim to show Russia/USSR neither as the 'goody' nor as the 'baddy' but focus attention on the characteristics particular to it.

The Second World is biased insofar as its choice of titles and authors consistently refuses the bureaucratized scholarship and paralytic tongue which has characterized much of the Soviet writing. In other words, it will prefer authors who have shown originality and courage of content and form.

Western perceptions of the Soviet scholarly achievement, especially of its social self-analysis, have been usually negative in the extreme. This was often enough justifiable. Heavy censorship stopped or biased much Soviet research and publication. 'Purges' have destroyed many of the best Soviet scholars, with whole disciplines closed on orders from above. The Soviet establishment has excelled in the promotion of safe scholars – the more unimaginative and obedient, the faster many made it into the limelight. However, much of the hostile detachment of the Anglo-Saxon scholarship and the media orginated in its own ideological bias, linguistic incompetence and a deeper still layer of mutual miscomprehension. To understand the human experience and thought in a particular social world, one must view it on its own terms – that is, with full awareness of its context – of history, political experience, culture and symbolic meanings. This necessitates the overcoming of stereotypes rooted in one's own experience and a challenge to that most persistent prejudice of all – the belief that everybody (and everything) is naturally 'like us', but somewhat less so (and that the best future mankind can have is to be like us but even more so).

The bafflement of the mainstream of Western scholarship at the dawn of Gorbachev's reforms has accentuated the collective miscomprehensions of Soviet society. On the one hand stand those who see nothing happening because nothing can happen: 'totalitarianism' is not open to any transformation from within. On the other hand stand those to whom the USSR is at long last becoming 'like us'. Both views miss the most important point, that Soviet society is moving along its own trajectory which can be understood only on its own terms. This makes the need to allow this society and its scholars to speak to us in their own voice, an urgent one.

Uniformity and uniformization are false as perceptions of history and wrong as social goals, but so also is any effort at keeping human worlds apart. This is true for international politics, scholarly endeavour and daily life. Half a century ago a Soviet diplomat, Maxim Litvinov, a survivor of the revolutionary generation which was then going under, addressed the League of Nations to say: 'Peace is indivisible'. The World War to follow did not falsify this statement, but amended it. Peace proved divisible but only at the heavy price of human peril. The same holds for knowledge.

Teodor Shanin
University of Manchester

Introduction: Alexander Chayanov as a Theoretician of the Co-operative Movement

By Viktor Danilov

Russia had arrived at the 1917 Revolution with a rapidly growing co-operative movement and with the widely accepted view that co-operatives had an important role to play in the country's future. Many people then believed – and with reason – that the co-operative movement would offer to Russian society ways of overcoming the social difficulties which inevitably accompany economic modernization rooted in industrialization. What seemed particularly important for Russia's agrarian society was the opportunity of involvement in a market economy through the extension of co-operation to an enormous mass of small peasant households. Indeed, over a period of some fifteen years the country had seen the growth of a broad network of consumer, credit, agricultural, craft, trade and other types of co-operative. By the beginning of 1902, a total of 1,625 co-operative associations had been registered in Russia: by the beginning of 1912 the numbers were 18,023; and by the beginning of 1915 they had reached 35,200. Their membership, according to approximate calculations, comprised between 11 and 12 million households. Given that a peasant family had on average 5 to 6 members (and it was the rural type of co-operative which decisively predominated) this meant that up to 60 million people, that is one third of the population of the Russian Empire,[1] were directly drawn into the sphere of influence of the co-operative movement.

In Russia, as in all other countries, the growth of co-operation did not come about simply 'from below', spontaneously or of its own accord. From the very beginning, the most important factors in its growth were public awareness, the active role of the progressive intelligentsia and – after the revolution of 1905–7 – state support for the creation and development of a system of small-scale co-operative credit. It seemed that an answer had at last been found to the most agonizing question of the post-reform period: how to rescue the peasant population of Russia from ruin and proletarianization – from the destruction of the peasantry as a class. Between the 1860s and the 1880s, the *narodniki* (populist movement) had hoped, with the assistance of an egalitarian type of land commune, to 'keep capitalism out' of the Russian countryside and lead the countryside directly to socialism. Following the example of Robert Owen, they tried to create socialist communes based on politically conscious members of the intelligentsia and farming partnerships of poor peasants. But these naive hopes did not survive the ordeals of life.

The initial Marxist critique of the *narodnik* illusions was a persuasive one: the commodity-capitalist development of the Russian economy was inevitable; and it occurred as an objective historical process which could not be prevented either by the commune or by the peasant partnership (*artel'*). Large-scale production based on machinery created new opportunities for the development of society, for the enhancement of its material well-being and culture and for the social emancipation of Man. But at the same time, this kind of production played a powerful role in dictating the transition to a market economy and swept out of its path the natural-patriarchal forms of small-scale production which typified the peasant and craft economies of Russia. The appearance and growth of the proletariat became the main social factor determining the present and future of the country. This was the interpretation from which the Russian Marxists started out, linking their hopes with the proletariat in political struggle for socialism.

However, this common interpretation of the historical process did not entail any unified political strategy or course of action amongst the different schools of Russian Marxism – nor, in particular, between its two main tendencies, the Bolshevik and the Menshevik. The Social-Democratic (Menshevik) tendency directed its political activity mainly into parliamentary channels. It put a great deal of effort into work in trade union, co-operative and cultural-educational organizations. It was from among the Mensheviks that many of the Soviet co-operative leaders emerged, including the head of the Central Association of Consumer Co-operatives in the 1920s, L. M. Khimchuk. (For Lenin, the name of this co-operative activist became a term of description: 'Khimchuk is useful because he knows how to create shops', 'the Khimchuks are doing useful work' and so on.)[2] The Bolsheviks did not refuse to undertake practical work in the co-operative system and other 'legal' organizations. However, they regarded this work as something of secondary importance which represented, all in all, merely an attempt to adapt to existing conditions – whereas the main task was to achieve a radical change in existing conditions, through a political and social revolution. This attitude, however, substantially limited the influence of the Bolsheviks on the co-operative movement – both as Marxist theoreticians and as political leaders; and it did so not only during the pre-revolutionary period but also following the 1917 Revolution. So far as Lenin was concerned, it took the entire experience of revolution to enable him fully to appreciate the merits and potential of the co-operative system.

For both the Bolsheviks and the Mensheviks, moreover, the field of co-operative activity was confined before 1917 to consumer societies of industrial workers, since in co-operatives of other kinds there was practically no participation by workers. Furthermore, the peasant and craftsmen's associations were trying to defend what seemed outdated forms of small-scale production – an activity which, in the eyes of the Marxists of that time, seemed to be, if not reprehensible at all events useless. This dogmatic mistake did a good deal of harm and caused a good many difficulties during the period after the Revolution.

By the beginning of the twentieth century, the crisis of Russia's small-scale production both in industry and in agriculture became generally recognized. But, only in the last analysis was it reflected in the growth of the working class; a much more widespread and obvious result of this crisis was the mass pauperization of the population: the non-proletarian impoverishment of the working people, which is familiar in developing societies today. In these conditions, for socialists the defence of small-scale production not only ceased to be useless, but became indispensable. The possibility of fairly effective solutions to this problem by means of co-operatives were by now known and had been tested by the experience of other countries. The figures quoted above on the growth of Russian co-operative associations between 1902 and 1915 demonstrate how strong was the public demand for a social mechanism to protect the consumer and small producer within the market economy, and the extent to which the country had, from this point of view, become ripe for the development of co-operation.

Russian social thought had been powerfully influenced by the idea of co-operation. An enormous number of books, pamphlets and articles in newspapers and journals had been published which actively propagated the ideas of co-operation and described the organization and functioning of the most varied kinds of co-operative associations, together with their economic and social achievements in Britain, Germany, France and other countries. Among the relevant publications of that time, there was hardly a book or article to be found which did not refer to the ideas of Robert Owen as 'the spiritual father of co-operation', or mention the principles of 'the Rochdale society', or the creator of peasant co-operatives F. Raifeizen, the followers of Fourier, the Fabians or other founders of the co-operative movement in the West.

Amidst the torrent of propagandist literature, as well as the descriptive or educational literature, serious works of scientific analysis began to appear relating to the organization of co-operation in Russia and to the theory of co-operation. Among these, a special place undoubtedly belongs to the books by S. N. Prokopovich and M. I. Tugan-Baranovskii, two well-known Russian economists of the beginning of the century. A book by the first-named author appeared in 1913 and provided an analytical picture of the initial stages of development of all forms of co-operative in Russia; and contained at the same time a detailed critique of how they were interpreted by contemporaries. The combination of a serious analysis of actual social and economic processes with a sharply polemical manner of exposition was in general characteristic of S. N. Prokopovich's works; and was also apparent in his book on the theory and practice of the co-operative movement in Russia.

The very definition of co-operation of Prokopovich contained a polemical edge – directed against the collectivist illusions which still survived among the first Russian co-operative activists. For Prokopovich, any co-operative association represented a free and self-managing alliance of members enjoying full and equal rights. It 'does not swallow up the individuality' of its members, but, on the contrary, 'offers full scope for their individual tastes and gifts', 'performing economic operations' relating to production exchange and credit 'on behalf of the members as a whole', but on condition that net income is distributed 'in proportion to the extent to which each member participates in the common work' (and not in proportion to the share capital). It existed for the purpose of increasing the productivity, and income from the work, of its members, for the purpose of lightening the burden and reducing the costs of their 'production and of the management of their households', which would in the long run make it possible 'to free them from the exploitation of the middleman, the shopkeeper and the money-lender and also make large-scale capitalist production unnecessary'.[3]

According to this definition, co-operation was not a universal form of alliance. Prokopovich makes this quite clear: 'The only people who can participate in co-operative organizations are those with some economic asset', 'only the economically prosperous elements'.[4]

Those who were not economically better off remained outside the ambit of co-operatives. The various types of partnerships of peasants, artisans and workers were regarded as remnants of outdated epochs and formations.[5] The modern co-operative system, according to Prokopovich, functioned solely under the conditions of the market economy and in accordance with its laws.

Whilst noting the 'dependence of co-operative forms on economic relationships', Prokopovich came to the conclusion that 'co-operation cannot serve as a weapon in the struggle against the advance of capitalism'.[6] Only 'for its own members' did it become an 'instrument of self-defence against the exploitation of capital' – by raising the productivity of, and income from, labour and by ensuring the 'inexpensive acquisition' of necessary products. It was specifically within these limits that co-operatives would wage a struggle 'against the exploitation of working people by the *representatives* of financial, commodity and productive capital', by turning it from 'the owner of the enterprise . . . into a hired factor of production'.[7]

This conceptual model was completed by an interesting theoretical argument which was formulated in Marxist language: 'On the basis of one and the same mode of production it is possible for totally different social ways of life to develop'.[8] The question therefore was of the formation of a 'co-operative way of life' within the framework of a capitalist economy. And apart from the argument as to who would ultimately be 'hired' by whom as a factor of production, this version of co-operative development was realistic. It was precisely this form which was, by and large, implemented in modern countries, where a capitalist economy includes a significant co-operative sector.

It was within the framework of market co-operation that Prokopovich also set out to solve the problems of agricultural co-operation. Life experience (such as the development of 'steam-powered transport', the growth of taxes, the penetration of market relationships into the countryside and so on) had already confronted the peasants with the task of 'restructuring their economy on new principles', the essence of which lay in the transition 'from the old type of natural economy designed to satisfy the needs of the peasant family to a new, money economy working for the market and making use of the services of the market'.[9] But the market inevitably meant 'the appearance of the trader and middleman' as an extremely egoistic intermediary between the peasant and the market. It was precisely this which necessitated a 'combination of the peasants' in co-operatives which would pursue 'both agricultural and broader economic goals'.[10]

A concrete examination of the development and functioning of different forms of agricultural co-operation – in relation to marketing, supply, reprocessing and, in particular, butter manufacture and credit – enabled Prokopovich to make some extremely pertinent observations concerning the process of the implementation of co-operation in the countryside. He demonstrated first of all that 'the natural character of our peasant economy' and 'the inadequate development of money relationships' within it, represented 'the main obstacle in our country to the development of agricultural co-operation' as a whole. Secondly and for that reason, the process of implementing co-operation involved 'only certain branches of the peasant economy' and this was itself reflected in the regional specialization of co-operative work.[11]

It is important to note that the difficulties of the co-operative movement in Russia could not be ascribed solely to objective conditions connected with the level of economic development. S. N. Prokopovich pointed very specifically to the administrative and legal barriers (above all to the fact that co-operatives required official sanction). 'The bureaucratic regime which is dominant in our country', so he wrote, 'is extremely antagonistic towards the principles of collective self-determination, of which co-operation is one particular form'.[12] His book ends with the assertion that the first precondition for the development of Russian society has become 'the winning of certain political rights – including the right of all citizens to combine freely in co-operatives and engage in autonomous activity'.[13]

The second of the authors referred to was M. I. Tugan-Baranovskii. A 'legal Marxist' and socialist who belonged to no political party, he devoted a major book, *The Social Foundations of Co-operation* (*Sotsial'nye osnovy kooperatsii*, 1916) to the general theory of the co-operative movement and to a generalized account of the development of all its forms and tendencies, not only in Russia but also in other countries. This book had been preceded by an earlier one – *Towards a Better Future* (*K luchshemu budushchemu*, 1912) – which was a first attempt to present such a theory. What we are therefore presented with is not a set of piecemeal arguments on a routine topic, but the fruit of serious study carried out over many years. It is important to emphasize this because the book might almost have been specially written for the Revolution; it contained what was virtually a premonition of the agonizing search for the path to a new society, and of the ruinous mistakes and irreparable losses which accompanied this search. We shall see below how, in the country where this book was written and published, it was only towards the end of the 1980s that 'socialism' was able to grasp what N. I. Bukharin (and, of course, not he alone) had begun to grasp in the 1920s; and yet it was something, so it now turns out, which had already been said in 1916 – on the eve of the Revolution.

Co-operation, according to Tugan-Baranovskii, was a new form of economic organization which had arisen 'as a result of the conscious efforts of broad social groups to transform the existing [i.e. capitalist – V.D.] economic system in a certain direction'. This 'direction' was that of 'the socialist ideal'.[14] Its implementation, even under capitalist conditions, had led people to combine in co-operatives – in economic enterprises whose practical activity differed in no way from that of capitalist enterprises, since they were pursuing 'the private economic advantage of their members' and were doing so 'through the medium of exchange'. A co-operative 'emerges fully equipped with capitalist technology, it stands on capitalist ground and this is what distinguishes it in principle from socialist communes which sought to create an economic organization on an entirely new economic basis'.[15]

Socialist communes which were set up in the nineteenth century in various countries, were opposed to capitalism by virtue of their very design, and required the initial existence of 'a new Man' or 'of people of exceptional moral qualities'. This in itself limited their influence and ultimately doomed them to failure. Tugan-Baranovskii considered it beyond dispute that although 'such communes are fully capable of surviving by their own efforts under favourable conditions' nevertheless their economies 'not only fail to provide their participants with the enormous advantages of which their sponsors had dreamt but are, as a general rule, worth scarcely more than an economy run by a single individual'.[16] A co-operative, on the other hand, must 'trust an individual as he is and must take the social environment as it is'. It 'builds something new . . . out of the raw material provided by contemporary society'.[17] This novelty is reflected above all in a cardinal change in the 'social-economic nature' of the economic enterprise.

Tugan-Baranovskii perceived the 'non-capitalist nature' of a co-operative enterprise in the fact that it 'never pursues the goal of earning a capitalist profit' although it does make a payment for the (share and loan) capital which it attracts. In a *consumer association* payment was made 'at the lowest possible rate of interest per share whilst the total net proceeds are distributed between those who collect the goods, the consumers'. However these proceeds 'do not in general constitute income in the economic sense. They represent only that part of the expenditure which has been saved by the members.' Likewise, in the case of a marketing association, the income was distributed 'not in proportion to the share capital' (except in the sense that the lowest possible interest rates were paid on the capital), but in accordance with the quantity of products which were 'made available for marketing' and which had been produced by the association's own labour. Lastly, in the case of a *producers' partnership* – 'the whole purpose' consisted in the attempt 'to eliminate the capitalist owner by handing everything over to the workers themselves . . . whilst the surplus product which, in the hands of the enterpreneur represented profit, will, by virtue of remaining in the hands of the workers who created the product, become earned income'.[18] Thanks to all this, the co-operative became a form of 'self-defence by the labouring classes against encroachment by the hirers of labour'.[19]

Up to this point, Tugan-Baranovskii's analysis of co-operation did not basically differ from that of Prokopovich. One may suppose that the possibility, as co-operation developed, of the emergence of a multi-layered economy based on the capitalist mode of production was something which Tugan-Baranovskii did not rule out – at least as a particular stage in the historical process. But he did not stop at that. In his view, co-operation represented not only a means of self-defence for the labouring population, but a breakthrough into a future socialist society.

Already before the Revolution, the 'legal' (as opposed to the 'actual') Marxists had grasped something which others would take a very long time to grasp: a socialist economy will inevitably continue to be based on commodity-money relations, since at this stage of historical development it is only the means of production which are socialized, whilst articles of consumption pass 'into private ownership so as to avoid encroaching on private life, which presupposes the right to choose what one consumes' and this is just the same 'as in the case of present-day commodity production'.[20] This is all the more important in view of the inevitability of the continuance of hired labour, since 'even in a socialist state the workers would not be the owners of the goods which they produce as well as of the means of production' and would have to be paid 'a definite wage for their labour' just as happens in the case of co-operative enterprises.[21]

Consumer co-operation, at least insofar as it involved the proletariat, constituted a ready-made part of the socialist economy; and if it extended to the entire population of the country and to the management of the national economy as a whole, it would represent nothing other than a system of collectivism, that is of socialism. Proletarian consumer co-operation, so Tugan-Baranovskii maintained, would tend towards the complete transformation of the existing social structure and towards the creation of a new economy based on 'the subordination of the entire economic system to the interests of social consumption'; it would 'gravitate towards collectivism'.[22] He understood however that the power of the co-operatives themselves was inadequate, and that they represented only a part of the workers' movement – existing side by side with trade union organizations and political parties.

The socialist potential of the co-operative movement was, according to Tugan-Baranovskii, confined solely to workers' consumer co-operatives. Not only did he deny that there was any socialist potential in petty bourgeois co-operatives (consisting of officials, for example) whose functions were confined to no more than those of adaptation to market conditions. He also denied any socialist potential in peasant or agricultural co-operatives. Their purpose was solely to defend the interests of 'the small peasant farmer' – the interests of 'his self-preservation', since capitalist conditions constituted a threat to 'the entire economic existence of the peasant as an independent farmer'.[23]

Co-operatives, so Tugan-Baranovskii believed, did a great deal to defend and improve the peasant economy and even did a great deal 'for its profound transformation'. However, this transformation was confined to the organization and expansion of market relations, the surmounting of the prevalent isolation of the small-scale peasant household and its involvement 'within a powerful web of social ties'. He recognised that 'the new type of peasant economy which is being created by co-operation' was becoming 'socially regulated'; it was being given 'the opportunity to make use of the gains and advantages of a large-scale economy' and 'to compete against large-scale capitalist enterprises'.[24] But nevertheless, the idea of transforming the peasant economy itself and of organizing its output on socialist principles was categorically rejected. This was partly connected with the fact that in all countries, agricultural co-operation embraced 'mainly the middle and rich peasants. The least well-off peasants, who are close to the proletariat, are economically too weak to participate in co-operatives in significant numbers.'[25] This, as Tugan-Baranovskii rightly observed, may explain the conservative, often plainly reactionary political attitudes of agricultural co-operatives in Germany, Belgium and France at that time.[26] But the main thing, so he maintained, was that co-operatives defended and strengthened the position of the peasant precisely as a small-scale individual producer, and did not create any collective economy. Let us take, for example, the following argument:

> The peasant household, even when drawn into co-operative organizations, continues to be a small-scale entity in the sense that the co-operative fabric – no matter how many ramifications it may develop and no matter how stable or complex it may become – is still based on the individual peasant household headed by an independent peasant farmer who manages the

farm at his own risk and peril. Not only does co-operation represent no threat to the independence of the peasant household: it makes the peasant household more secure by making it more successful and by improving its technical standards. One must therefore reject, in the most categorical fashion, any idea that co-operation leads to the concentration of the peasant economy or thereby prepares the ground for socialism[27]

Only in a producers' partnership did the possibility exist for 'the complete absorption of individual agricultural production by social production', but neither in agriculture nor indeed in industry, as Tugan-Baranovskii noted, was this 'in practice' widespread at the time of the book's publication.[28]

It is strange to read such a defense of what came to be called collectivization in the writing of a 'legal' Marxist who inclined to a position which was reformist and not in the least revolutionary. The dogmatic interpretation of the organization and functioning of especially large-scale advanced production solely in the image and likeness of factory or machine-powered production, based on the collective labour of large groups of workers, was at that time almost universal – as was the simple equation of socialism with collectivism. We should emphasize, however, that Tugan-Baranovskii's theoretical models were in no sense practical recommendations. On the contrary, when we study them we find the following clarification:

> contemporary socialism in no sense requires the destruction of peasant ownership or the replacement of the labour of the peasant on his field by the tilling of the land by large social groups. Most representatives of contemporary socialism . . . recognise that even within the framework of a socialist state, the peasant household may be preserved, owing to the substantially different conditions of production in agriculture and in industry. By not making demands for the destruction of peasant ownership, socialists may also appear as defenders of peasant co-operation which undoubtedly supports the peasant economy.[29]

One begins to understand why Tugan-Baranovskii, although he denied that there was any socialist potential in peasant co-operation, was nevertheless able, as a socialist, to advocate its further development. Co-operation as a whole was 'a creative, constructive force'; it contained 'a spirit which draws mankind onto new paths and which creates new social forms'. Moreover it was already 'building a new society within the framework of the one which now exists'.[30] Such were the general ideas of Tugan-Baranovskii's book. In a country which was becoming ripe for social revolution, his general theory of co-operation, permeated as it was with optimism, was actually perceived as a scientific validation for the co-operative development of Russian society.

* * *

The subsequent progress of co-operative theory in Russia is linked with the name of Alexander Chayanov. A scholar, writer and active public figure, Chayanov perished during Stalin's repressions and has only recently been rediscovered: first in the West (in the 1960s) and subsequently in his native Russia (though only at the end of the 1980s). Chayanov's book *The Theory of Peasant Economy* (*Teoriya krest'yanskoi ekonomiki*) appeared in an English translation in two editions – in 1966 and 1986–7. The introductory articles to the book which were written by D. Thorner, B. Kerblay, R. Smith and T. Shanin provide a good picture of the author's career and contribution to learning.[31] These articles have concentrated mainly on the theory of the peasant economy, which was the main object of Chayanov's concern and of his thinking as a scholar and public figure. The present book represented his basic work on peasant co-operation – which not only makes it possible to modernize, and therefore rejuvenate and rescue the peasant economy of family farmers, and enhance the well-being and culture of the countryside, but which also makes it possible to restructure the life of society as a whole, including rural society, advancing social justice and general prosperity.

The publication of works devoted to the modernizing of agriculture marked the beginning of Chayanov's life as a scholar. As a 20-year-old student of the Moscow Agricultural Institute (now called the Timiryazev Agricultural Academy), Chayanov at the end of his second year of study in 1908 spent his vacation in Italy and then, in 1909, in Belgium. In both cases his vacations were taken up with studying the work of co-operatives in the countryside and the agronomical services which they provided. The choice of these subjects was not in itself out of the ordinary: they were at that time the 'sore spots' of Russian society. Moreover, his teachers had included the eminent expert on agronomy, D. N. Pryanishnikov and the well-known economist A. F. Fortunatov, who had themselves made extensive studies of both agronomy and rural co-operation. Fortunatov had been the author of one of the first courses on co-operation for students in Russia.[32] Pryanishnikov in 1904 had visited

Italy and had drawn attention to the important role of agronomical services in the development of its agricultural co-operatives.

There was nothing fortuitous nor was there anything new in studying the experience of other countries which had achieved significant successes in agriculture by means of an efficient agronomical service and of a widely developed system of co-operatives. The beginning of the century in Russia had been marked by quite a number of such studies. By way of example we should mention the writings, which became well-known, of Professor A. N. Antsyferov – the author of *Co-operation in the agriculture of Germany and France* (*Kooperatsiya v sel'skom khozyaistve Germanii i Frantsii*), which was published in 1907; this author also delivered a series of lectures at the Shanyavskii University on 'The present state of credit co-operatives in the countries of Western Europe', which was published in 1913.

Chayanov's student writings proved to be quite out of the ordinary. This was recognized at the time, as is shown by the fact that they were published at once in the mainstream press of that day. The central All-Russian journal *Vestnik sel'skogo khozyaistva* ('The Agricultural Gazette') had, already in 1908 printed in two of its issues Chayanov's article on the work of agronomists in the Italian countryside. In 1909, his 'Letters from the Belgian countryside' appeared in 6 issues of this journal; and in 1910, spread over 15 issues was his analysis of Belgian data concerning agricultural credit, 'public measures' relating to cattle-rearing and the insurance of cattle. In 1909 a pamphlet by Chayanov followed under the title *Co-operation in Italian Agriculture* (*Kooperatsiya v sel'skom khozyaistve Italii*). Of course these letters and essays devoted a great deal of space to descriptive material on matters of special relevance for practical work in Russian agriculture. However, they were written in a lively and persuasive way and involved a quest for those aspects of advanced experience in other countries which could be adopted and utilized in Russian agriculture in order to help solve its own problems, above all for the improvement and advancement of the small-scale peasant economy. It was precisely this emphasis on the social aspects of research, and the highlighting of general questions concerning the importance of co-operation, organization and agronomy which constituted the distinctive characteristics of Chayanov's early writings. They had outlined a range of problems which would ever afterwards remain crucial in his scientific activity.

In his pamphlet on peasant co-operation in Italy he had focused attention on elucidating the role of co-operation in the general and rapid advance of agriculture which could be seen in that country at the end of the last century and the beginning of the present one.

Here he came across genuine evidence of the capacity of co-operatives not only to defend the peasant against the offensive from private capital, particularly the money-lender and the middleman, but also of its capacity to create an economic mechanism to underpin the adaptation of the peasant economy to market conditions, thus promoting a general advancement. Italy's growing agricultural production. so Chayanov maintained, 'was not artificially contrived in isolation from the popular masses . . . but was, on the contrary, engendered from the depths of the national economy. Since it marked a real economic renaissance of the entire nation, it contributed powerfully to the enhancement of the well-being of the popular masses.'.[33]

What seemed to Chayanov to be particularly valuable in Italian experience was the direct link between the work of credit co-operatives and that of publicly organized facilities with regard to agronomy. This, in the first place, guaranteed the rational and efficient organization of credit facilities for peasant households (in Russia this problem was still unsolved). Secondly, this system made it possible to carry the financing of agriculture beyond the stage of partial, often piecemeal, improvements and to proceed to the stage of influencing regional and national production as a whole. In this way, new conditions were created for the growth of co-operatives and for the enhancement of their role. Direct comparisons between those cases where a credit system was established and those cases where it was not (as in Russia) – provided extremely eloquent testimony in favour of co-operation.[34]

In the case of Belgium, too, Chayanov was particularly interested in the interaction between co-operatives, publicly organized agronomical facilities and state policy, in the process of the radical 'marketization' of agriculture, and in the sharpening conflict between large-scale and small-scale forms of production in which the peasantry found itself under attack. It was while he was studying Belgian co-operatives and peasant households that Chayanov arrived at a new interpretation of both. This interpretation was prompted by the high level of efficiency, organization and rationality in the structure of the peasant co-operative system which made the component elements and connections in this system transparent. The conclusion Chayanov formulated for the first time stated that the analysis of the internal structure of the peasant economy 'reveals to us a number of distinct technical processes in agriculture which are, through the deliberate activity of "economic man" integrated into an "economy" . . . Each one of these processes is so self-contained that it can be detached in a technical sense wihout disturbing the general organizational plan of the economy'.[35]

This conclusion very soon developed into a formulation of the main theme of Chayanov's research: the theory of the peasant economy. It was precisely here, in his essays on Belgium, that one can discern the starting-point of his explanation of the importance of peasant co-operation which he described as 'the possibility – without making any special changes in the economic equilibrium and without substantially destroying the organizational plan of the small-scale rural economy – of organizing some of its particular technical economic activities where large-scale production enjoys an undoubted advantage; organizing these activities up to the level of large-scale production by technically detaching them and merging them with similar activities being undertaken by neighbours, into a co-operative'.[36] The words just quoted were constantly to recur in his writings right up to the end. The idea which they expressed was developed and added to, it was to be reinforced by fresh arguments and fresh investigations and meanings; and it was to develop into a general theory of peasant co-operation which was most fully expounded in the present book.

The young author of the 1909 articles understood that if peasant households were to be brought within the co-operative system, conditions of commodity production were essential. 'Such a merger' so he observed '. . . does not take place until processes are introduced of a kind which can be merged; and it is particularly important to remember this in Russia' (where the involvement of peasant households in commodity relations was only just beginning). But once these conditions existed, then in his view, the process of bringing peasant households into the co-operative system became a matter of objective necessity. The circumstances in Belgium (the crisis of small-scale farming and so on) 'compelled small-scale farms to organize certain aspects of their organizational plans along co-operative lines – by dint of the same historical necessity whereby the development of machinery had led to large-scale capitalist production in industry'.[37] This comparison of the importance of the extension of co-operatives with that of industrialization does, of course, contain an element of exaggeration. At that time, however, there had been no development of non co-operative methods of introducing elements of large-scale production and distribution into farms of the peasant type.

A key had been found to the solution of one of the main problems of the social-economic development of Russia – the problem of the peasants' road into modern society. However, it would require years of further analytical work and – what was particularly important – of practical work in the Russian co-operative movement, before the understanding of the organizational structure of co-operatives could evolve into a theory of co-operative development for the peasant economy and into a programme of practical action.

Already at the stage when Chayanov was becoming established as a scholar and public figure (and it is in this way that the prerevolutionary years of his career can be defined), he was showing a characteristic desire and ability to connect science with life and to turn knowledge into practice. He rapidly became an active figure in the co-operative movement in Russia, and was one of the organizers and leaders of the Central Association of Flax-Producers' Co-operatives (*L'notsentr*, which was set up in 1915). It was natural that his research and teaching activity should have combined an analysis of practical questions (his publications of those years included articles on the co-operative marketing of agricultural products, the co-operative study of markets, the work of agronomists in co-operatives, the teaching to peasants of co-operative book-keeping, the setting up of co-operative associations and so on) with an analysis of questions of a conceptual nature, involving, in particular, the introduction of a general teaching course on co-operation, based on lectures delivered at different places.

A. V. Chayanov's *Short Course on Co-operation* (*Kratkii kurs kooperatsii*) was published in 1915 and enjoyed great popularity. In the years 1919–1925 the book was to be re-issued in three further editions. (Several further editions were to appear after his rehabilitation in 1987.) In its treatment of the general questions of co-operation, this course comes close to the conceptions of M. I. Tugan-Baranovskii; but in its interpretation of peasant co-operation it differs very substantially from those conceptions and even contradicts them; since it regards the peasant household as the basis of a future agrarian system – perfected both in its productive and in its social aspects; and it regards co-operation as the path to the creation of such a system and as a form of its existence.

The formation of Chayanov's theory of the co-operative development of the peasant economy, and of the political programme connected with it, was decisively influenced by the Russian Revolution of 1917 which opened up opportunities for a choice in the path of social development and which provided experience of social and economic transformations. From the very first days of the Revolution, Chayanov was an active participant who fought for the implementation of profound democratic and social reforms, including radical transformations in agriculture. In April 1917 he was one of the founders of the League for Agrarian Reforms, whose aim was to bring about a discussion of the agrarian question and ways of solving it. As a member of the Central Managing Committee of the League, he took part in drawing up a document whose purpose was 'to define the parameters within which ... a discussion of the agrarian problem should take place. We are of the opinion (1) that the self-employed co-operative peasant farm should form the foundation of the agrarian system in Russia; and that our country's land should be handed over to it; (2) that this transfer should take place on the basis of a state plan for land organization, drawn up with due regard for the special features of the life and economy of different regions and implemented in a planned and organized way without damaging the productive effort of our national economy; (3) that land organization is only a part of the solution to the agrarian problem, which involves all matters connected with the general conditions of agricultural production, the organization of self-employed peasant farms and the organization of links between these farms and the world economy as a whole'.[38]

Apart from A. V. Chayanov, the members of the Managing Committee of the League included some of the most eminent agrarian specialists of that time, who, moreover, represented different political tendencies (N. P. Makarov, P. P. Maslov, S. L. Maslov, N. P. Oganovskii and others). The platform presented in their joint names was evidence of the fact that the idea of agrarian reform orientated towards a co-operative peasant economy had become widespread and had gained wide support from public opinion in the country. The validation of this idea was the subject of Chayanov's brilliantly written pamphlet *What is the Agrarian Question?* (*Chto takoye agrarnyi vopros?*) (It was printed as the first in a series of publications by the League on basic questions of agrarian reform). The textual similarities between Chayanov's pamphlet and the three points of the platform quoted above indicate that they were written by the same hand. These two documents taken together do indeed represent a political programme for the solution of the agricultural question in Russia, drawn up by Chayanov and supported by the League for Agrarian Reforms.

The starting point for Chayanov's programme included the most revolutionary demands of 1917, and above all the demand which had become the slogan of all democratic forces: 'The land – to the working people!'. 'In accordance with this demand', so the pamphlet explained, 'all land now forming part of the farms of large landed estates must be handed over to self-employed peasant farms'.[39] This 'transfer of privately owned land to the peasantry' could be carried out in the form of socialization, in the sense of the abolition of any ownership of land ('it belongs equally to everybody, like the light and the air'); or in the form of nationalization, that is, the transfer of the land into the ownership and control of the state; or in a form involving a decisive role for local self-management in the control of the land, and involving the use of a 'single tax on land' in order to collect a ground rent for the benefit of the people (following the idea of Henry George); or finally, through the creation of a 'system of state regulation of land ownership' with a ban on the right to buy and sell land.[40]

None of these alternative solutions to the problem was totally excluded: the choice was eventually to be made by the Revolution. However, Chayanov himself sought to find the best solution to the complex problem of 'implementing the socialization of the land and its transfer to the self-employed peasant farms with the minimum difficulties and the minimum costs'. He was inclined to favour a combination of the last two alternatives – a system of state regulation of land ownership and a system of progressive land taxes with the additional right 'to expropriate any land', since he believed that this would make it entirely possible 'automatically within one or two decades to achieve nationalization or municipalization'.[41]

The socialization of land ownership and its transfer to the peasantry were only the starting point of an agrarian reform whose aim was infinitely broader and more important, namely to create 'a new agrarian structure and a new kind of land ownership'.[42] The purpose of the reform was to ensure 'the development of productive forces' and the creation of 'new production relations' of a kind which would meet two basic criteria (or, what amounted to the same thing, two principles of a particular conception of organization and production), namely (1) *the maximum productivity of the labour invested by the people in the land*; and (2) *the democratization of the distribution of the national income*.[43]

When assessing, in the light of these criteria, the '*conceivable* systems of production relations', Chayanov rejected not only capitalism but also 'state socialism' and 'anarchistic communism' although it seemed to him at that time that they might be regarded as 'theoretical organizational forms' of the principle of the democratization of distribution. However the task, in his opinion, consisted in 'bringing both organizational principles into harmony with one another'.[44] Only co-operation was capable of achieving this and therefore the future agrarian system must be based on co-operatives.

The case for a co-operative future now began to rest not only on the possibility of protecting the small-scale peasant household under conditions of market competition, but also on the economic and social merits of co-operation in comparison with those of a large-scale capitalist economy. Chayanov saw the real advantages of the latter in market specialization and in the use of complex machinery and of the achievements of science ('the availability of agronomists', 'improved cattle' and so on). Relying on the yardsticks used at that time, Chayanov considered, first, that 'the very nature of an agricultural enterprise places limits on the enlargement of its scale' and therefore, that 'the advantages of a large-scale over a small-scale economy in agriculture could never be very great in quantitative terms'. Secondly, practical experience led him to the conclusion that co-operation had the capacity 'to impart all these advantages of a large-scale economy to small-scale peasant households'. Moreover 'small-scale peasant households, when joined in co-operative associations, achieve a scale and potential which is actually greater than those of the very largest private farms'.[45]

This comparison between small-scale and large-scale farms was not confined to the sphere of production. Chayanov was led by the logic of his thinking to an analysis of social differences: 'We have to compare, not large-scale and small-scale farms, but a farm which is operated by its owner and by the manpower of the owner's family, on the one hand; and, on the other hand, a capitalist farm which is operated by hired labour'.[46] For Chayanov, this description in itself contained the proof of the advantages of the peasant household, advantages which were to be revealed to their full extent under the conditions of a co-operative system. In those conditions the path was opened to the constant intensification of labour, the growth of production and social wealth combined with a guarantee of democratization in the distribution of the national income.

In 1917, as a 'non-party socialist' who was also the author of a socialist and radical programme of agrarian reform, Chayanov rapidly won political authority. He became a member of the Main Land Committee, which had the task of politically supervising the preparation and later implementation of land reforms; he became a member of the Council of the Russian Republic which exercised supreme state power pending the convening of the Constituent Assembly; and finally, he was appointed assistant to the Minister of Agriculture in the last Provisional Government.[47] There is every reason to suppose that agrarian reform in conditions of further democratic development would have borrowed many of the ideas set out in Chayanov's programme. However the pursuit of such an agrarian reform proved impossible owing to the narrow mindedness of the ruling classes, the weakness of democratic institutions which had not yet been fully formed, and the shortsightedness and political blindness of the people who led these institutions.

In the Russia of 1917–9 a peasant revolution had broken out which had destroyed ownership by landlords as well as private land ownership in general. It had taken the form of the direct seizure and redistribution of land; and had no connection whatever with co-operatives or with any other formerly defined programme. The wave of this revolution led to the success of the Bolshevik workers' revolution, which placed the Communist Party headed by Lenin in power; they, in turn, embarked on revolutionary socialist transformations implemented by dictatorial methods. A political dictatorship of the working class was established which very soon degenerated into the dictatorship of the Communist Party and later, by the end of the 1920s, degenerated into Stalin's bureaucratic dictatorship. But, in the years 1917, 1918 and 1919, a popular revolution was in progress which was deciding the fate of the country.

A. V. Chayanov belonged to the stratum of the Russian intelligentsia which, despite its extremely complex, contradictory and mainly negative attitude towards the October Revolution of 1917, very soon began to work within Soviet institutions or within cooperative organizations which were collaborating with Soviet power; and did so with obvious and increasing benefit to the Soviet Republic. Here we can quote the testimony of Chayanov himself. At the beginning of 1930 when he was being subjected to sharply intensifying persecution, he managed to publish in the *Sel'skokhozyaistvennaya gazeta* ('Agricultural gazette') an article entitled 'On the fate of the neo-*narodniki*' which was written with astonishing sincerity, almost as if he had already anticipated his own fate. This is what he then wrote about his attitude to the October Revolution: 'In general *I entirely agree with the view once expressed by Jaurès that a revolution can be either completely rejected or equally completely accepted, just as it is.* I have been guided by this view throughout all the years since our revolution took place, Therefore the question of my attitude to the October Revolution was decided not at the present time, but on that day in January 1918 when the Revolution discarded the idea of the Constituent Assembly and followed the path of the proletarian dictatorship. Ever since February 1918, my life has been bound up with the revolutionary reconstruction of our country; and, as I carefully recall, day by day, the years which have passed, I believe that no one has, or can have, any grounds for refusing to describe me as a Soviet worker, without any inverted commas'.[48]

The complexity of Chayonov's relations with the Bolshevik leadership lay in the fact that whilst he joined in the common endeavour, he maintained his independent views and openly criticized everything in Bolshevik policy which he considered to be incorrect, mistaken, harmful or simply unnecessary from the point of view of socialist policy or from the point of view of the tasks and opportunities of the period of transition to socialism. Chayanov's standpoint was openly and precisely formulated in his review of Nikolai Bukharin's book *The Economics of the Transitional Period* (*Ekonomika perekhodnogo perioda*, 1920). The gist of Chayanov's observations was that '. . . many of the phenomena of disintegration in our national economy do not spring endemically from the transitional period [as they were portrayed in this book – V.D.] but are the inevitable result of measures which have not been thought through, which are unnecessary or not obligatory . . .'.[49] This was precisely the standpoint of Chayanov which has emerged from recently published documents of co-operative conferences and congresses in 1919 and in the following years.[50]

Chayanov's main pronouncement during the period of the Bolshevik Revolution and civil war was his book entitled *The Basic ideas and organizational forms of peasant co-operation* (*Osnovnye idei i formy organizatsii krest'yanskoi kooperatsii*) which appeared in 1919 and was an early version of the book now published in a shortened form in English translation. The writing and publication of the monograph was the result, above all, of a decade of research work which made it possible to provide a concrete picture of the organization and functioning of all forms of agricultural co-operation and of all the main branches of its work – and thus to provide an extended validation for the concept of the co-operative peasant economy on the basis of enriched and reinforced arguments. There was something else which was no less important: the book represented a direct answer to the questions raised by the course of the Russian Revolution. It is well-known that from the autumn of 1918, priority was being given in Soviet agrarian policy to the creation of collective and Soviet (i.e. state) farms. This first experiment in collectivization did not, and could not, lead to serious results. On the contrary, it provoked widespread opposition among the peasantry and, in the spring of 1919, the goal of collectivization was dropped for the purpose of practical politics, although it was retained in socialist policy programmes as a long-term perspective which remained constant, (although requiring the creation of the objective preconditions, strict adherence to the principle of voluntary membership and so on). The collective farm movement, which mainly included some revolutionary elements of the landless strata in the countryside, enjoyed total support from the state. All the greater therefore was the urgency of a critical analysis of the nature of the experiment in collectivization which had taken place. The book, which appeared in that same year, 1919, directly in the train of events, decisively rejected 'the communization of production' in agriculture as well as 'the co-operative socialization of the entire peasant economy'; and pointed out such genuinely difficult problems as 'labour incentives', 'the organization of labour' and 'the managerial will', that is, the issues of management.[51] The experience of subsequent Soviet history has made us very familiar with the difficulty of solving all these problems. However the idea of collective agriculture (based on *artel'*, that is partnerships) was in no way totally discarded: it became an integral part of the general concept of the co-operative development of the countryside.

A specific familiarity with the organizational-productive structure of the peasant economy and of the possibility of separating it into parts and of splitting off certain productive and organizational functions, made it possible, as A. V. Chayanov showed '. . . to split off and organize in the form of large-scale co-operative enterprises, those sectors where such an enlargement of scale would produce a noticeable positive effect . . . without disturbing those aspects of the economy where small-scale family production was technically more convenient than large-scale production'. In the end, the opportunity was created for organizing all levels of activity, all functions and types of work, etc., 'on the particular scale and on the social foundations which are most appropriate for them'. Thus, side by side with the peasant household, a 'large-scale collective enterprise of the co-operative type' was arising.[52] In 1919 it was commonplace to emphasize the subordinate role of co-operation in relation to the peasant economy. This was reflected in the following definition: 'Peasant co-operation . . . is a part of the peasant economy which has been split off for the purpose of being organized on large-scale principles'.[53] Co-operation would exist for as long as the peasant economy existed; and it was this which predetermined 'the limits of co-operative collectivization'.[54]

Only a few people were able to understand the originality and depth of the idea of 'co-operative collectivization', particularly in the conditions of 1919–1920. But there is no doubt that the idea of 'co-operative collectivization' found endorsement in Lenin's article 'On co-operation', of 1923, particularly in the article's conclusion that for the Russian peasants the growth of co-operation was in itself identical with the growth of socialism. It is known that Chayanov's book was one of the seven books on the theory and practice of co-operation which Lenin consulted when he dictated the article on his death bed.[55]

The second edition of the book on peasant co-operation was extensively revised and supplemented – as stated on its title-page. The preface to the book, was dated 1 December 1926 – which indicates that the changes and additions were not prompted by political expediency, nor were they due to coercion nor in general were they dictated by extraneous factors (which was to become commonplace in publications on public affairs from the beginning of 1928). The New Economic Policy, orientated towards the socialist development of the countryside through co-operation, as well as the practical experience of co-operation itself in the 1920s, had provided a genuinely new, rich and important factual material for verification, clarification, the perfecting of the system and from a conceptual point of view. The theoretical discussions which took place in the 1920s on economic and social problems provided a great deal of food for thought.

What was entirely new in comparison with the first edition of the book was an analysis in depth of the problem of the 'horizontal' and 'vertical' types of concentration of production, of their potential and of the interrelationship between them. The importance of this theoretical analysis has been confirmed by the experience of agricultural development in different countries. Either there is a powerful upsurge of production, accompanied by social progress, along the path of 'vertical' concentration, that is, along the path of growing diversity and interaction between different forms and scales of the organization of production processes and economic ties, both of the co-operative and the non-co-operative varieties. Or by contrast, production will stagnate and there will be social stalemate, if the path of 'horizontal' concentration is followed – assuming, of course, that this is not accompanied by something much worse (i.e. brutal coercion as happened under Stalin's collectivization). From the point of view of socialist development, 'vertical' integration in its co-operative form was obviously to be preferred. The book indicated how, in the long term, the establishment of a co-operative system of agricultural production means that '. . . the entire system undergoes a qualitative transformation from a system of peasant households where co-operation covers certain branches of their economy into a system based on a public co-operative rural economy, built on the foundation of the socialization of capital which leaves the implementation of certain processes to the private households of its members, who perform the work more or less as a technical assignment.'[56]

'Horizontal' concentration in the form of collective farms was in no way rejected out of hand. Collective farms, set up by peasants of their own free will – on their own initiative and in their own interests – could and should be part of the co-operative system in accordance with general co-operative principles. 'The choice' so Chayanov wrote 'would not be between *collectives* and *co-operatives*. The essence of the choice would be whether the membership of co-operatives is to be drawn from collectives or from peasant family households'.[57] He did not exclude the possibility (although he did not think it the most desirable) of a 'concentration' of production embracing the whole of agriculture, in which 'literally all peasant households were ultimately merged into communes and were organized on optimum areas of 300 to 500 hectares' [741 to 1235 acres]. But it was emphasized that this should 'in no way affect our basic system of co-operatives engaged in purchasing, credit, marketing and production, which would continue to be organized as before. The only difference would be that the membership of primary co-operatives, instead of being drawn from small peasant households, would be drawn from communes'.[58]

'Co-operative collectivization', so A. V. Chayanov believed, represented the best, and perhaps the only possible way of introducing into the peasant economy 'elements of a large-scale economy, of industrialization and of state planning'.[59] What he saw as its merit was that it was implemented on an entirely voluntary and economic basis which amounted to 'self-collectivization'.

The idea of 'co-operative collectivization' reflected the basic tendency of the actual development of co-operation in the Russian countryside in the 1920s, and offered a real alternative to collectivization of the Stalinist variety. This was quite enough to ensure that the book would very soon be condemned and banned; and that its author would be among the first victims of Stalinist repression.

NOTES

1. See S. N. Prokopovich, *Kooperativnoye dvizhenie v Rossii, ego teoriya i praktika*, (*The Co-operative Movement in Russia, its Theory and Practice*), Moscow 1913, A. V. Merkulov, *Istoricheskii ocherk potrebitel'skoi kooperatsii v Rossii* (*A Historical Outline of Consumer Co-operation in Russia*), Petrograd, 1915.
2. V. I. Lenin, *Complete Works*, 5th Russian edition, Vol. 37, pp. 230–231, 232 et al.
3. Prokopovich, *supra*, pp. 16–17.
4. Ibid., pp. 10, 11.
5. Ibid., pp. 21–29.
6. Ibid., p. 29.
7. Ibid., pp. 15, 30. My italics – V.D.
8. Ibid., p. 30. The idea of the possibility of forming and developing different social structures on the basis of one and the same mode of production was a new idea which ran counter to prevailing Marxist assumptions of the day. It is gaining scientific currency in our own time.
9. Ibid., pp. 114, 115.
10. Ibid.
11. Ibid., p. 121, 153 et al.
12. Ibid., p. 444.
13. Ibid., p. 453.
14. M. I. Tugan-Baranovskii, *Sotsial'nye osnovy kooperatsii* (*The Social Foundations of Co-operation*), Berlin, 1921, p. 4. We have made use of the copy of the book which exists in the library of Cambridge University (Great Britain).

15. Ibid., pp. 68, 71.
16. Ibid., pp. 67, 78.
17. Ibid., p. 67.
18. Ibid., pp. 76–77, 29, 87.
19. Ibid.
20. Ibid., pp. 7, 9.
21. Ibid., Ibid., p. 74.
22. Ibid., pp. 498–499.
23. Ibid., pp. 346, 351.
24. Ibid., pp. 361, 368–369.
25. Ibid., p. 349.
26. Ibid., pp. 349–360.
27. Ibid., p. 363. Cf. pp. 499–501.
28. Ibid., p. 366.
29. Ibid., pp. 367–368.
30. Ibid., pp. 96, 105 et al.
31. See: A. V. Chayanov, *The Theory of Peasant Economy*, Ed. by D. Thorner – Homewood, 1966; *A. V. Chayanov on the Theory of Peasant Economy*. Edited and introduced by D. Thorner, B. Kerbley and R. E. F. Smith. Second edition, with an additional foreword by T. Shanin. Madison, Wisconsin, 1986; Manchester 1987.
32. See A. F. Fortunatov, *Ob izuchenii kooperatsii. Kursy po kooperatsii*, (*On the Study of Co-operation. Courses on Co-operation*), Vol. III. Shanyavskii University Publishing House, Moscow, 1913.
33. A. V. Chayanov, *Kooperatsiya v sel'skom khozyaistve Italii*, (*Co-operation in the Agriculture of Italy*), Moscow, 1909, p. 4.
34. Ibid., pp. 6–7.
35. A. V. Chayanov, 'Letters on Belgian Agriculture', *Vestnik sel'skogo khozyaistva*, 1909, No. 36, pp. 8–9.
36. Ibid.
37. Ibid.
38. See the 'Preface' on behalf of the Managing Committee of the League of Agrarian Reform to its publications. (For example, A. V. Chayanov, *Chto takoye agrarnyi vopros?* ('What is the Agrarian Question?'), Moscow, 1917, p. 4; *Osnovnye idei resheniya agrarnogo voprosa* ('Basic Ideas for Solving the Agrarian Question'), Moscow, 1918, pp. 3–4 et al.
39. A. V. Chayanov, *Chto takoye agrarnyi vorpos?*, supra, p. 20.
40. Ibid., pp. 32–33, 41–45.
41. Ibid., pp. 55, 58–59.
42. Ibid., p. 38.
43. Ibid., pp. 16, 18.
44. Ibid., pp. 18–19.
45. Ibid., pp. 24, 25.
46. Ibid., p. 26.
47. The following is the description of A. V. Chayanov, which appeared on the list of candidates nominated to the Constituent Assembly by the national organization of the co-operative movement of Russia:
    '*Chayanov, Aleksandr Vasil'yevich*. Member of the Council of the

Russian Republic, representing co-operatives. Member of the All-Russian council of co-operative congresses. Teacher at the Petrov-Razumovskii Agricultural Academy. Member of the Main Committee on Land issues. Assistant to the Minister of Agriculture. Co-operative movement activist. Member of the Council of the Central Association of Flax Producers in Moscow. An economist. Well-known for his writings on the land question, the peasant economy and co-operation. In politics – a non-party socialist'. (*Golos naroda*, organ of the co-operative alliances and associations, 12 (25) November, 1917). It should be noted that Chayanov could not take up his duties in the Ministry of Agriculture.

48. *Sel'skokhozyaistvennaya gazeta*, 29 January 1930.
49. Quoted from N. Bukharin and G. Pyatakov 'The Cavalry Charge and the Heavy Artillery (A light-hearted response to the critics of 'The Economies of the Transitional Period)'. *Krasnaya nov'*. Literary-artistic and scientific, social-political journal. 1921, No. 1, p. 272. The reply tendered apologies to Chayanov for the use of his unpublished manuscript (p. 256). The manuscript has unfortunately not been found.
50. See, *Kooperativno-kolkhoznoye stroitel'stvo v SSSR. 1917–1922. Dokumenty i materialy*. (*The Building of Co-operatives and Collective Farms in the USSR. 1917–1922. Documents and Material*). Editor-in-chief, V. P. Danilov, Moscow 1990, pp. 133, 159, 162, 177, 180, 181, 278, 281, 301, 302, 304–306 et al.
51. A. V. Chayanov, *Osnovnye idei i formy organizatsii krest'yanskoi kooperatsii* (*The Basic Ideas and Organizational Forms of Peasant Co-operation*), Moscow, 1919, pp. 42, 301, 303–305.
52. Ibid., pp. 15–16.
53. Ibid., p. 21.
54. Ibid., p. 301.
55. See V. I. Lenin, *Complete Works* 5th Russian edition, Vol. 45, pp. 376, 597–598.
56. A. V. Chayanov, *Osnovnye idei i formy organizatsii sel'skohozyaistvennoi kooperatsii* (The Basic Ideas and Organizational Forms of Agricultural Co-operation), 2nd ed, further revised and supplemented, Moscow, 1927, p. 13. See p. 11, this book.
57. Ibid., p. 345 [Russian]; this book, p. 205.
58. Ibid., pp. 341–342 [Russian]; this book, p. 204–5.
59. Ibid., p. 24 [Russian]; this book, p. 21.

Author's Comment

This inquiry is based on the personal experience of the author – who has worked for some twenty years in the ranks of the Russian co-operative movement – and on his observation of the co-operative movements in Italy, Belgium, Germany, Switzerland and France. It is also based on the work of a seminar on questions of agricultural co-operation which has been running for many years and which has been led by the author since 1913 at the Timiryazev (formerly the Petrov) Agricultural Academy.

The book includes substantial revisions of its first edition, which have been undertaken so as to keep pace with developments in co-operative theory. It also includes, among other things, material originally presented in a report to the co-operative section of the congresses of the Supreme Council of the National Economy (VSNKh) in 1919.

The author has deliberately narrowed the scope of his inquiry to an analysis of the basic ideas of agricultural co-operation and to an investigation of its basic organizational forms. We have not attempted to describe fully the present state of agricultural co-operation, still less have we attempted to outline its history, since we had neither enough material nor enough time for such an ambitious inquiry.

Yet even within the limited scope of our subject, we are still far from having gained a complete or final grasp of the material because the focus of our study – the peasant co-operative movement – is developing so rapidly in its scope and depth that theoretical ideas have lagged behind its practical achievements. Hence the many defects of this book and, perhaps, the element of haste in its conclusions, which seek, ahead of time, to provide a logical framework for the new forms of a spontaneous movement which are now being historically envisaged.

Colleagues who read the page-proofs of this book drew my attention to the not entirely clear way in which I had used the terms 'state capitalism' and 'capital' in relation to the peasant household. I therefore think it necessary to make it clear in the preface that:

1. In relation to the peasant household, which has no variable capital, I have used the term 'capital' in its most general sense; and it does not, of course, have the historical connotation associated with the capital within capitalist production.
2. In exactly the same way, I have used the term 'state capitalism' in the sense in which it was understood in our country in 1923: that is, as a synonym for a planned economy, based on state enterprises employing hired labour. No special social or political connotation is implied in this term.

The Author

Petrovsko-Razumovskoye\
1 December 1926

1

The Processes and the Concept of Vertical Concentration in the Rural Economy: Peasant Co-operation as an Alternative

It is highly likely that a great many of the readers of this book – agronomists, engineers, teachers and those who work in the rural community – have more than once despaired when faced with the obstacles that the life of a modern Russian village has placed in the way of what they are doing.

It must be realized, of course, that there are good reasons for this. No one will deny that the basic idea underlying the organization of the contemporary economy is the idea of large-scale organizational measures, involving many thousands of workers, tens of millions of roubles of capital, gigantic technical constructions and the mass production of standardized goods. The Ford machine-building factories, the *Volkhovstroi* and the other gigantic hydroelectric plants, the ocean-going transport lines, which are serviced by high-powered transatlantic enterprises, banking concerns that concentrate milliards of roubles of capital into a powerful economic fist – these are the economic facts which dominate and fascinate the minds of those who manage the economy at the present time.

It is no wonder, therefore, that many of our comrades, especially our younger comrades, whose minds are still full of images of the goals and achievements of the present-day industrial economy, and who are impatient to achieve something similar in their own provinces, are often reduced to utter despondency after a few months' work. They come close to desperation as they get jolted in a peasant cart on a rainy November evening along impassable roads from somewhere like Znamensk via Buzayevo to somewhere like Uspensk. Everywhere they encounter a lack of roads as well as the poverty and indifference of peasants installed on small, overlapping agricultural allotments, and who, with an excessive 'purely petty-bourgeois stupidity', are turned inwards on their tiny holdings.

We are sympathetic to the desperation of such a comrade who has visions of being in a Ford workshop although in reality his work obliges him to deal with two people farming five acres of ploughed land with one cow and often without a horse.

However, while understanding the subjective desperation of the beginner, we are in an objective sense totally disinclined to accept his depressing conclusions.

It is still quite obvious, of course, that the economic life of the peasant countries – China, India, the Soviet Union and many other countries of Eastern Europe and Asia – does not provide us with the sort of visible and obvious achievements of new organizational forms which we can easily see in the industrial countries of the West.

However, any economic phenomenon should always be examined from the point of view of the way it evolves, and it should, so far as possible, be examined in depth. If such an approach is adopted towards the agriculture of peasant countries, it then turns out, greatly to the surprise of many people, that agriculture is not only not hopelessly unsuited to the application of wide-ranging organizational policies, but that it is precisely here, in our present epoch, that intensive changes are occurring. These changes are making it into a subject whose organizational scope is of no less importance than the large-scale innovations in industry. It would therefore be useful in the highest degree for our despondent reader to understand that it is precisely in those outlying areas that the greatest potential exists for the goals and achievements of the future.

The point is that, so far, these changes are only at their very first stage of development and that they cannot and do not obviously provide us yet with anything like a complete picture which can be made the subject of a photograph.

The whole purpose of this book is to show the paths along which our countryside is developing and the forms in which it is being organized. As a result, there occurs a barely perceptible, but in reality a most radical, restructuring of the countryside's organizational foundations. The countryside, which only ten or twenty years ago presented an anarchical picture of diffuse, minute, semi-cultivated households, is now on its way to becoming the object of the most wide-ranging structural innovations and the base for large-scale economic activities.

If, on the basis of the evidence, we study the historical course of the development of urban industry and banking, we can easily see that their modern forms of organization, whose power and scale so impress us, have certainly not always existed. They are the result of gradual and in fact fairly recent development. In the not so distant past, some 150 years ago, the textile and even the metal-working industries were organized in the form of small-scale craft enterprises, often run as family concerns. And it was only capitalism, as it developed and grew stronger, which caused the disintegration of the patriarchal forms of the organization of production on the basis of crafts; and which, after first gaining control over the trade turnover, created the first large enterprises in the form of textile mills, subsequently multiplied them in the form of modern factories and plants and, in the final phase of its development, consolidated them into trusts and syndicates of various kinds.

There is no need for us to describe this process in detail: it will be known to our readers from any textbook of political economy.[1] For the purpose of the present book, the one important thing about this process of capitalist development is that in the sphere of agriculture this process was retarded; and that in many places its evolution assumed somewhat different forms.

There is, of course, no doubt that in agriculture, as well as in industry, large-scale forms of economic organization yielded considerable advantages and lowered production costs. In agriculture, however, these advantages were not as apparent as in industry.

The reason for this lay in the technical conditions of agricultural production. Indeed, the main form of concentrating and enlarging the scale of production in industry was that of so-called *horizontal concentration*, i.e. a form of concentration under which a multitude of very small and geographically scattered enterprises were merged, not only in the economic but in the technical sense, into one gigantic whole which concentrated enormous reserves of manpower and mechanical power into a small space and thereby achieved a colossal drop in the production costs. But in agriculture the achievement of such a degree of horizontal concentration was unthinkable.

What is meant by agriculture is basically the utilization by humans of the solar energy that reaches the earth's surface. The solar rays that radiate on to 100 acres of land cannot be focused onto 1 acre of land. They can be absorbed by the chlorophyl of the crops sown only over the entire radiated area. Agriculture is, by its very nature, inseparably connected with a large area; and the greater the scale of an agricultural enterprise from the technical point of view, the larger the area that it has to occupy. In that sense, no concentration in space can possibly be achieved.

Let me quote a small example. A factory owner who possesses a 100-horsepower engine and who wants to increase his output tenfold can install a 1,000-horsepower engine and thereby considerably reduce his working costs. But let us suppose that a farmer who tills his strip of land with one horse wants to achieve a tenfold increase in his sowings. He naturally cannot acquire a horse that is ten times larger: he has to acquire ten horses of as good a quality as the first horse. There will be some reduction of costs through a substitution of tractor power for horse power. But the farmer who already has one tractor and who increases his sowings tenfold cannot increase the power of the tractor: he has to acquire ten similar machines to work simultaneously in different areas, as a result of which working costs will be reduced to a considerable extent. The same can be said with respect to other kinds of stock – seed, manure, cattle, and so forth.

A farmer, when he increases his production, is in most cases obliged to increase the number of the things that he uses, rather than to increase their scale. Because of this, the economies of scale, expressed in quantitative terms, are smaller. It should furthermore be noted that the very nature of agricultural production imposes a natural limit on the enlargement of an agricultural enterprise.

Granted, then, that agriculture is inevitably diffused in space, it follows that a farmer has to move an enormous number of objects around within this space. Horses and animals have to be moved around; so also do machines, manure and finished products.

The larger the household, the greater the land area it works. Therefore, the greater will be the quantity of the products and the greater the distances over which they will be transported; and therefore there will be a constant increase in the costs of transport within the household, both in relation to its economic activity as a whole and per unit of finished product.

The more intensively the economy develops, the more deeply and thoroughly the land will be ploughed up and the greater will be the use of manure. Therefore, the more frequent will be the expeditions from the farmstead into the fields and the greater will be the burden of these journeys to and fro on the costs of production.

With the grain economy based on extensive farming operating in the Orenburg or Samara provinces, the proprietor needs to make only two expeditions: to sow the crops and to gather them in. But as soon as he begins to undertake an autumn ploughing for the sake of spring crops and to cart manure onto the fields, the number of expeditions increases many times over, as we can observe in our central agricultural provinces. Any further resort to intensive methods – the replacement of cereals by beet, turnips or potatoes – will increase the number of journeys to such an extent that every additional increase in the distance between the fields and the farmstead will make itself felt.

The entire benefit derived from the enlargement of the scale of production may be swallowed up by a rise in the cost of transport within the household, and the more intensive the household becomes, the more rapidly will this swallowing up occur. In Orenburg and Samara, our peasant households often extend over areas of 2–3,000 *desyatiny* [5,400 to 8,100 acres] run from one farmstead. In the Voronezh province, after the transition to three-field systems, the size of the optimal unit of exploitation fell to 800 *desyatiny* [2,160 acres]. In the Poltava province such an enlargement would already have been impossible. In the Kiev province and in Western Europe the costs of transport within the household will still further reduce the area of the household, bringing them to optimal levels of 200 to 250 *desyatiny* [540 to 675 acres].

It not infrequently happened that in earlier times, when households began to be managed by more intensive methods, large-scale owners were obliged to split up their estates into a number of separate farmsteads. Although they were large landowners, they were small or medium-scale land cultivators.

Thus the very nature of an agricultural enterprise imposes limits on its enlargement; and therefore, in quantitative terms, the advantages of a large household over a small one in agriculture can never be very great.

So, despite the fact that even in agriculture large-scale forms of production had an undoubted advantage over small-scale forms, we must nevertheless recognize that in *quantitative* terms these advantages were by no means as significant as in manufacturing industry.

Owing to the fact that, from a quantitative point of view, the advantages of a large-scale economy were less than in the case of industry, the peasant households could not be so simply or decisively crushed by the large latifundia, as their counterparts, the family craftsmen, were crushed by the factories. Furthermore, the peasant households demonstrated an exceptional capacity for resistance and tenacity of life. While often starving in the difficult years, working to their utmost capacity, sometimes recruiting hired labour and thus themselves assuming a semi-capitalist nature, they held firm almost everywhere and in some places even extended their land-holding at the expense of large-scale capitalist agriculture. The wave of post-war agrarian revolutions which swept through Eastern Europe and even gripped Mexico strengthened their position still further.

However, the fact that the peasant economy demonstrated such a great capacity for survival in the universal economic struggle for existence did not in itself in any way mean that it was to remain untouched by the general capitalist development of the world economy. Capitalism, owing to the technical conditions which we have described, was unable to organize agriculture according to the principles of horizontal concentration and invariably looked for other ways of gaining control over the anarchy of agriculture and of organizing it according to capitalist principles. Instead of the not very well-suited forms of horizontal concentration, the gaining of control took the form of *vertical concentration*.

In actual fact, the most recent research into the development of capitalism in agriculture indicates that the involvement of agriculture in the general system of capitalism is in no way bound to take the form of the creation of very large farming enterprises organized on capitalist lines and operating on the basis of hired labour. Repeating the stages of the development of industrial capitalism, agriculture, having emerged from forms of semi-natural existence, was subordinated to commercial capitalism. This in its turn – sometimes by means of very large commercial enterprises – drew into its sphere of influence large numbers of scattered peasant households. By gaining control over the links between these small commodity producers and the market, commercial capitalism subordinated them to its economic influence; and, by developing a system of credit on conditions amounting to slavery, turned the organization of agricultural production into what was a special kind of exploitative distribution system based on squeezing the workers dry. In this connection one should recall those types of capitalist exploitation which the Moscow cotton firm 'Knopa' used in relation to cotton-growers: buying up their harvest in the spring; handing out advances for food; and granting credit in the form of seeds and tools of production. Such a commercial firm, being interested in the standardization of the commodity which it was buying up, would quite often also start to interfere in the organization of production itself, imposing its own technical standards, handing out seed and manure, laying down what the crop rotation was to be, and turning its clients into technical executors of its schemes and of its economic plan. A characteristic example of this kind of arrangement in our country was the planting of beet in peasant fields under contracts with sugar factories or contractors.

Having gained control over outlets to the market and having created for itself the base for raw material, rural capitalism begins to penetrate the production process itself, splitting certain sectors of activity away from the peasant household's activities, mainly in the sphere of the primary reprocessing of agricultural raw material as well as sectors connected with mechanical processes. The entrepreneurial, steam-powered threshing machines, which travelled round southern Russia offering services for hire, the small butter factories in Siberia at the end of the nineteenth century, the flax-processing workshops in Flanders and in some parts of our own flax-cultivating provinces – all these are vivid examples.

If, in relation to the most developed capitalist countries such as those in North America, for example, we add to all this the extensive development of mortgage credit, the financing of working capital for households and the commanding power of capital invested in transport, elevators, irrigation and other enterprises, then we begin to see new ways by which capitalism penetrates agriculture. It turns the farmer into a source of manpower working with means of production belonging to others; and it turns agriculture, despite its apparent diffusion and the autonomy of its small commodity producers, into an economic system controlled on capitalist principles by a number of very large enterprises, which in turn are under the control of the highest forms of finance capitalism. It is no accident that, according to the calculations of Professor N. Makarov, out of the revenue from farming realized on the wholesale commodity exchanges in America, only 35 per cent goes to the farmers, while the remaining 65 per cent is absorbed by capital involved in railways, elevators, irrigation, finance and commerce.

In relation to this vertical capitalist concentration, the transformation of households from an area of 10 hectares [24.7 acres] to areas of 100 to 500 hectares [247 to 1235 acres], and the parallel transformation of a significant number of farmers from a semi-proletarian to an obviously proletarian position, would seem like a minor phenomenon. And if this phenomenon is absent, the evident reason is that capitalist exploitation brings a higher dividend precisely through vertical rather than horizontal concentration. In this way, moreover, it transfers a significant part of the entrepreneurial risk from the owner to the farmer.

The form of concentration of agricultural production just described is characteristic of almost all young agricultural countries which are engaged in the mass production of products of one type intended for remote, mainly export, markets.

Sometimes, as a result of the situation that has arisen in the national economy, this vertical concentration assumes forms which are not capitalist but co-operative or mixed. In this case, control over the system of enterprises involved in commerce, the handling of elevators, land improvement, credit and the reprocessing of raw material, which concentrate and manage the process of agricultural production, belongs wholly or in part not to the owners of capital but to small-scale commodity producers. They have organized themselves and have invested their private funds in enterprises or have succeeded in establishing social capital for that purpose.

The appearance and development of co-operative elements in the process of the vertical concentration of agriculture becomes possible only at certain stages of the process itself; these co-operative elements can appear only when local capital is relatively weak. Here, we deliberately emphasize the word 'relative', since this relative weakness of local capitalist entrepreneurs can result not only from their own weakness in absolute terms but can also result either from the prosperity of the peasant household itself (as in Denmark) or from the fact that co-operative elements may be backed by the financial resources of the state or by the resources of large-scale foreign capital or industrial capital which needs unadulterated raw material.

A vivid example of this process is the development of butter manufacture co-operatives in Siberia. At the end of the nineteenth century, after the building of the great Siberian railway, there arose in western Siberia as the result of the abundance of land suitable for fodder, a situation that was extremely favourable to the development of butter manufacturing for export. In the area of Kurgan, Ishim and other districts, small-scale entrepreneurs began to appear one after another. They soon covered the area with small factories producing butter, thus beginning in a capitalist form the process of the vertical concentration of agriculture in western Siberia.

Siberian butter manufacture, which had been created by small company promoters, in the course of decades turned an originally favourable market into one that was unprofitable. It came up against an acute crisis owing to the very large number of established factories and their ferocious competition, both for milk supplies and in the selling of their butter. Surviving thanks not so much to the revenue from butter as to the profits derived from the shops and payments in goods for the milk, these factories eked out a wretched existence for a number of years and then began, one after another, to close down. For the peasant households which had already re-organized themselves into market-oriented dairy farms, these closures entailed the threat of heavy losses. Not wishing to go back to a natural economy, they were faced with the question of whether to take over the factories which were closing down and run them according to the principles of a traditional peasant partnership (*artel'*).

The co-operative factories which eventually arose in this way were distinguished by the superior quality of their product over the adulterated butter produced by private entrepreneurs; and their development therefore received financial backing from the commercial capital of Danish and English export firms, which had their Siberian offices in Kurgan and in other cities. They rapidly squeezed the private entrepreneur out of the sphere of butter production.

In this way, the concentration of Siberian butter manufacture, which was begun by small-scale industrial capital, was continued with the support of large-scale commercial capital in co-operative forms. 'The Siberian alliance of butter manufacturing co-operatives' made an entry onto the London market and, by relying on bank credit, it totally rid itself of the influence of local commercial capital.

In several different forms, but in accordance with the same dynamics, and also after forming various kinds of connections at different times with capitalist groups, there also developed other kinds of agricultural co-operation.

What has just been said is more than enough for the purpose of understanding the substance of agricultural co-operation as a particular form of the thoroughgoing process of the vertical concentration of agriculture. It must, however, be noted that in the case of co-operative forms this process takes place at a much deeper level than in the case of capitalist forms, since in the case of co-operative forms of concentration, it is the peasant himself who hands over sectors of his economy which capitalism does not succeed in forcibly wresting from peasant households. Such is our understanding of the vertical concentration of agricultural production in the conditions of capitalist society – a concentration which extends both to purely capitalist and also to co-operative forms.

When looking through the statistics of co-operation we see that at the present time co-operative forms of vertical concentration of agriculture have reached an extremely impressive scale. Present-day agricultural co-operative organizations in our country number within their ranks millions of households, and their turnovers have long since been reckoned in hundreds of millions of roubles.

It is precisely this process of the extension of co-operatives to our countryside that we would like to draw to the attention of our despairing worker in the countryside – as the initial phase of the journey which alone can bring the agriculture of peasant countries to a complete and decisive re-organization on the basis of the most large-scale organizational measures.

The now observable forms in which this process manifests itself are modest and not very obvious. What, indeed, can be remarkable about the fact that a peasant woman, after milking her cow, washes her can and uses it to take milk to a dairy association in a neighbouring village? Or in the fact that a peasant who cultivates flax takes his fibre not to the bazaar but to a co-operative reception centre? But in fact, this peasant woman with her minutely small can of milk is linked to two million similar peasant women and peasants and forms part of the co-operative system of butter centres, which constitutes the largest dairy firm in the world and is already perceptibly re-organizing the entire structure of peasant households in the dairy farming regions. And the small flax cultivator, who already possesses sufficient staying-power as a co-operative member, represents one particle in the co-operative system of flax centres, which is one of the most important factors constituting the world market in flax.

This is happening at the very first stages of our movement. Its future prospects are incomparably vaster. However, when adopting a programme for the vertical concentration of agriculture in co-operative forms, we have to envisage that this process will last for a considerable time. Like the consecutive phases of capitalist development, from its initial forms of elementary commercial capitalism to the present-day factory, and system of trusts embracing the whole of industry – the vertical concentration which is developing in co-operative forms in our agriculture must inexorably pass through a number of consecutive phases of historical development.

Beginning as a rule with the combination of small-scale producers for the procurement of agricultural means of production, co-operatives very soon turn to the organization of the co-operative marketing of agricultural products which they develop in the form of gigantic alliances combining hundreds of thousands of small-scale producers. As the intermediary operations of this type acquire the necessary scope and stability, they form the basis for a smoothly functioning and powerful co-operative apparatus and, what is particularly important, there occurs, in a manner analogous to the development of capitalism, an initial accumulation of co-operative capital. During this phase of their development, agricultural co-operatives, under the pressure of market forces and as a matter of historical necessity, evolve into organizations with their own operations for marketing and for the primary reprocessing of agricultural raw materials (involving co-operatives for butter manufacture, potato grinding, canning, the dressing of flax, and so on); they remove the relevant sectors of activity from the peasant households; and, by industrializing the villages, they gain control of the commanding positions in the rural economy. Under our conditions, thanks to the assistance of the state and to the granting of state credits, these processes are being accelerated and may occur simultaneously and become interwoven one with another.

Having extended the co-operative system to marketing and technical reprocessing, agricultural co-operatives thereby bring about a concentration and organization of agricultural production in new and higher forms, obliging the small-scale producer to alter the organizational plan of his household in conformity with the policy of co-operative marketing and reprocessing, to improve his technology and to adopt more perfect methods of land cultivation and cattle-rearing, which ensure uniform standards for the product.

However, having achieved this, co-operatives inevitably build on this success with the aim of even more widely embracing the productive sectors of the peasant economy (by creating machinery users' associations, assembly points, associations concerned with the inspection and with the pedigree of cattle, with joint processing, land improvement, and so on); and part of the expenses involved in these kinds of co-operative production are covered, and must as a matter of principle be covered, out of the profits derived from marketing, procurement and credits.

Given the parallel development of electrification, of technical installations of all kinds, of the system of warehouse and public premises, of the network of improved roads and of co-operative credit – the elements of the social economy grow in quantitative terms to such an extent that the entire system undergoes a qualitative transformation from a system of peasant households where co-operation covers certain branches of their economy – into a system based on a public co-operative rural economy, built on the foundation of the socialization of capital which leaves the implementation of certain processes to the private households of its members, who perform the work more or less as a technical assignment.

Having carefully reflected on the enormous importance for agriculture of the process just described of vertical integration in its co-operative forms, we can maintain with conviction that from the point of view of the national economy the appearance of agricultural co-operatives is of no less importance than was the appearance of industrial capital a century earlier. We must note, however, that neither the importance of agricultural co-operatives nor, in general, the nature of the rural co-operative movement at the present time, have in any sense been adequately understood until now, even by their own creators and participants. This, incidentally, is quite understandable, since it is true of nearly all economic movements that theory comes considerably later than practice.

Capitalism, which is more than a century old, was in effect brought within the ambit of research only at the end of the last century and many of its most complex problems have not up till now been fully studied. Agricultural co-operatives represent such a young and as yet unformed movement that we have no right even to expect them to be the subject of comprehensive theoretical analysis. With few exceptions, all that we have for the moment is a co-operative ideology rather than a co-operative theory. Nevertheless, it is absolutely essential for us to ascertain in as much detail as possible the organizational forms within which, and the economic apparatus by means of which, agricultural co-operatives are carrying out and can carry out the colossal national economic functions which we have just described; under what conditions and under the pressure of what factors co-operative organizations come into being and are able to exist; and what incentives there are which set their energy in motion.

Without having a finalized or broadly perceived general theory of co-operatives, and making it our purpose to elucidate the nature of the new and gradually developing national economic system which is growing up on the basis of the vertical concentration of peasant households, we need first of all to ascertain what we mean by the concept of 'co-operation', i.e. the subject of study in this book. It means that we have to establish what the organizational and economic qualities are which we ascribe to the co-operative system discussed, and by what criteria we differentiate it from related formations of other kinds. In view of the widespread use in our everyday life of the word 'co-operation', this task might seem to be one of elementary simplicity. However, this is very far from being the case; and it should, perhaps, be recognized that this concept is among the most nebulous and unclear that we use.

It is commonly supposed – and this greatly obfuscates the real state of affairs – that our agricultural co-operative system is merely one variety of the general co-operative movement which also includes the system of consumer co-operatives in the towns as well as all kinds of co-operative associations of artisans and craftsmen. Their economic nature is deemed to be identical and, both in our country (in the *Mezhkoopsovet*, the inter-co-operative council) and in the International Co-operative Alliance, they are merged, even from an organizational point of view, into a single whole.

Despite this, or perhaps precisely because of it, we have not up till now produced any generally accepted formula which defines the general concept of 'co-operation'. And that remains true despite the repeated and numerous attempts of various authors and despite prolonged arguments over this question. Therefore, when setting out to clarify this term, we have to approach the question with particular caution; and we shall try above all to do so in positive terms by examining how the co-operative members and activists of the co-operative movement themselves define the substance of their organization, and what they consider to be, or not to be, 'co-operative'.

Over the past fifteen years, during which time this author has had occasion to talk to Russian, Belgian, Italian and German co-operative activists, he has heard the most diverse, and sometimes contradictory, views of what attributes represent the substance of the co-operative movement.

Some people maintained that the most important thing about co-operatives was that their membership was voluntary, that they were independent and that their management was democratic. Others attached importance to the methods by which profits were distributed and to the secondary role of capital in co-operative enterprises. A third group attached great importance to the openness of co-operative organizations, and thought accordingly that it would be contrary to co-operative principles to refuse to admit new members. A fourth group laid particular stress on the fact that co-operatives consisted of self-employed workers; and they were opposed not only to the admission of non-workers but also to the employment of hired labour in co-operative enterprises. A fifth group believed that the substance of co-operatives lay not in their organizational forms but in the social goals which they set themselves, i.e. in their struggle on behalf of the have-nots, which was in some cases socialist and in other cases religious in nature. A sixth group drew a distinction between co-operatives and communes on the basis that co-operatives involve only the partial socialization of economic activity and not the fusion of all economic effort into one collective enterprise. And so forth.

When we try to discern the meaning of the attributes just mentioned, and when we try to relate them with one another and with the phenomena of life, we are obliged to note that they are both variegated and contradictory. Many of these attributes are inapplicable to whole classes of co-operatives, whose co-operative nature is intuitively not open to doubt nevertheless.

Thus at the present time, for example, it is extremely hard to find a co-operative which does not employ hired labour. Also, the traditional craft associations (*arteli*), as well as many agricultural co-operatives (such as those concerned with land improvement, with machines and with land cultivation), very often consist of exclusive groups which do not admit new members. It is not uncommon to find consumers' associations where membership is obligatory for entire categories of staff. Finally, those who work in co-operatives do not by any means always set themselves social goals, or if they do so, the aims are often sharply contradictory – as we can see, for example, from the religious type of peasant co-operatives in Belgium and in the socialist workers' co-operatives. Therefore, if our overriding purpose is to provide a definitive formula applicable to co-operatives of all kinds, we have to include the common characteristics found in all branches and those that can be regarded as essential.

Proponents of the theory of co-operatives do precisely this – by providing extremely short and abstract formulas. Immediately before our revolution of 1917, enormous importance was attached to these definitive, general formulas: they became the subject of violent disputes; and in our own literature, dozens of formulations were put forward.

For the sake of exemplification and comparison we will take two of the most vividly expressed and most sharply opposed formulations. Thus, for example, Tugan-Baranovskii defined co-operatives as follows:

> A co-operative is an economic enterprise made up of several voluntarily associated individuals whose aim is not to obtain the maximum profit from the capital outlay but to increase the income derived from the work of its members, or to reduce the latter's expenditure, by means of common economic management. (M. Tugan-Baranovskii, *Sotsial'no-ekonomicheskaya priroda kooperatsii* [The Social and Economic Nature of Co-operation]).

The definition given by K. Pazhitnov sounds entirely different:

> A co-operative is a voluntary association of some individuals which aims, by its joint efforts, to combat the exploitation by capital and to improve the position of its members through the production, exchange and distribution of economic benefits, that is, as producers, consumers or sellers of manpower. (K. Pazhitnov, *Osnovy kooperatsii* [The Foundations of Co-operation]).

All the other formulations repeated or developed those of Tugan-Baranovskii and Pazhitnov, or else tried to combine their ideas into a single formula.

When comparing the defining characteristics quoted above, we can therefore easily divide them into two groups: on the one hand, those of an organizational and formal nature (the role of capital, methods of distributing profits, forms of management, etc.); and on the other hand, those relating to social goals (destruction of the capitalist system, class harmony, liberation of the peasantry from economic bonds, etc.).

The question naturally arises as to whether it is possible to combine these two categories into a single definitive formula. And doubt may even arise as to whether these definitive features are concerned with one and the same phenomenon. Indeed, we are profoundly convinced that the attempt to define co-operatives involves not one but two things which have to be defined. On the one hand, we have a co-operative enterprise as an organizational-economic entity which may not set itself any social goals whatever; or which may even set itself social goals which run counter to those contained in the formulations enumerated above. On the other hand, we see before our eyes a widespread co-operative movement, or, more exactly, co-operative *movements*, each of which has its own particular ideology and which makes use of co-operative forms for the organization of economic enterprises, as one of the instruments (sometimes the only one), for its concrete embodiment. These latter movements consciously set themselves social goals of various kinds, and would be unthinkable without such goals.

Therefore, in our opinion, the concept of 'co-operatives' has to be broken down into two concepts: a 'co-operative enterprise' and a 'co-operative movement', and defining characteristics need to be formulated for each of them.

A 'co-operative enterprise' can be quite adequately described by a formal definition of the kind given by Tugan-Baranovskii. One can, in any case, find several typical organizational elements (such as the role of capital or the working people's social environment) which make it possible to provide *a single definition* for all co-operatives.

From the formal, organizational-economic point of view, the Belgian co-operative of the religious type – the first paragraph of whose statutes lays down that membership is open only to those who recognize 'the family, property and the Church as the only foundations of society' – and the communist co-operative associations of workers may be totally identical.

But without trying to provide a brief and an all-embracing formula, we might nevertheless deem it to be a characteristic feature of a co-operative enterprise that it can never be a self-centred enterprise having its own interests existing apart from those of the members who set it up. It is an enterprise which serves the interests of its clients who are also its proprietors and who organize its management in such a way that it is directly responsible to them and to them alone.

All the elements of the definition given by Tugan-Baranovskii and those like him logically stem from the idea we have formulated; and it really can be a feature common to all co-operative enterprises. But such a uniformity is hardly possible once we turn to the characteristics of co-operatives as a social movement.

It is true that, at first, when co-operatives represent primarily a movement based on literature and on ideas, such uniformity does exist. But as soon as the co-operative movement penetrates the thick of the national economy and becomes one of its essential foundations, class divisions and other contradictions begin to appear and the ideological mirage vanishes into thin air. An attentive observer will note the separate workers' co-operatives which regard themselves as a part of the general workers' movement; the burghers' urban co-operatives; the artisan co-operatives; and finally, peasant co-operatives. Each of them, in so far as they have really become part of life and have put down roots, are of the same flesh and blood as those social classes and groups which gave birth to them. And if, within the given group, there arises any kind of consciously conceived social movement of a class nature, then co-operatives are inevitably used as one of the elements of such a movement. Thus, for example, the labour movement has been reflected in three concrete forms: in a workers' party, in a trade union and in workers' co-operatives.

In Western Europe in the peasant milieu something like this can also be observed (in Belgium, Switzerland *et al.*) where alongside agricultural co-operatives, which are sharply differentiated from and hostile to workers' co-operatives, there exist parties which base themselves on the economic interests of the peasantry. Given such a state of affairs, it would be naive in the conditions of a class society to regard all types of co-operative movement as parts of a single, unified whole, and to subsume them under the nebulous general concept of 'the struggle for the interests of the working people'. From the scientific point of view this would mean abandoning depth and detail in social analysis; and from the political point of view it would mean ignoring those sometimes antagonistic class interests which ought to be identified in all their different meanings.

Therefore, from the social point of view, we should always speak not about *the* co-operative movement but about co-operative movements. And we are profoundly convinced that the same distinction should also be drawn in the organizational analysis of co-operative enterprises as such. While acknowledging that it is possible to provide a single definition of a co-operative enterprise, if this is approached from the formal, administrative point of view, we must nevertheless emphasize that because of its generality, it is devoid of specific content and therefore virtually useless.

In particular, in relation to agricultural co-operation, the common definition of the formal-organizational type totally leaves on one side the organizational-economic content which is involved in co-operative work as the result of the process which we have described as the vertical concentration of agriculture which assumes co-operative forms. Yet it is precisely this process of gradual concentration and regeneration of agricultural production which represents the essence of the matter for us, the organizers of the new agriculture. And what is important to us about **agricultural producer co-operatives** is

The Processes and the Concept of Vertical Concentration 17

precisely those elements which distinguish it from consumer co-operatives, and in no way the formal elements which they have in common. Nor should we forget that the social nature of co-operatives inevitably affects the specific economic tasks of a co-operative enterprise and, therefore, its organizational goals.

The organizational and economic bias of a co-operative enterprise is influenced to an important degree by the role in production or in commodity circulation which is played by its owner-members. What, indeed, can there be in common between a private shop and a similar co-operative shop staffed by the disabled, involving exactly the same commodity and similar technique, and often operating in the same neighbourhood? Of course, from a formal point of view, both economic enterprises are similar. But from the point of view of political economy they are of a nature which is not only different but antagonistic.

There is an even greater difference of organizational tendencies with regard to co-operative warehouses, let us say, for fruit and vegetables, depending on whether they are maintained by co-operatives of consumers or those of producers.

It must never be forgotten that in the conditions of a society which is alien to the planned organization of the state economy and to the state regulation of production and of the market, a co-operative represents an element, organized on collective principles, of an economic activity of a group of individuals; and that its purpose is to serve the interests of this group and of this group alone.

A workers' consumer co-operative represents organized purchasing activity on the part of the proletariat which has no interests apart from the interests of the proletarian class. Craftsmen's associations which handle raw materials have a point and purpose only in so far as they supply craftsmen with material for work. Peasant co-operatives, as we know, constitute a part of the peasant economy which has been given a distinct identity in order to organize that economy on large-scale principles.

In short, then, co-operatives cannot be thought of in isolation from the social and economic foundations on which they are based; and in so far as these foundations are economically diverse, the forms of co-operatives are themselves of diverse kinds.

Co-operatives organize those interests and aspects of the lives of groups or classes which already existed before co-operatives appeared; and at the early stages of the development of capitalist society, co-operatives did not introduce any new element which stands above class relations.

Thus peasant co-operatives in our opinion represent, in a highly perfected form, a variation of the peasant economy which enables the small-scale commodity producer to detach from his plan of organization those elements of the plan in which a large-scale form of production has undoubted advantages over production on a small scale – and to do so without sacrificing his individuality. He is able to organize them jointly with his neighbours so as to attain this large-scale form of production – while possibly employing hired labour.

Urban consumer co-operatives involve neither such a combination of households nor the rationalization of their productive activity. Their purpose is to co-ordinate the purchasing activities of their members and to organize a more rational expenditure by members of the income which they get from productive activity outside the co-operative.

Partnerships of craftsmen, artisans and traders do not constitute such an association of independent economic units, but usually represent a complete fusion within one enterprise of the working activities of their members; they are, therefore, better described as joint production rather than co-operation.

In order to bring this idea as sharply as possible into relief, we would like to focus the attention of the reader on four specific scenarios.

Scenario A would assume an ordinary butter manufacturing association of peasants in which the process of manufacturing and selling butter has been split off from the households and organized into a co-operative which has built a factory and hired workers to manufacture the butter. The co-operative produces butter from the milk supplied by its members and sells it to consumers. It is in a state of antagonistic relations (in the sense of opposing interests) both with the consumers and with its own workers (who produce the butter).

Let us now suppose that our co-operative is obliged for some reason to go into liquidation and that the workers who produce the butter buy the butter manufacturing factory and organize themselves into a workers' butter-producing partnership, the economic nature of which can be seen in scenario B. Here the co-operative buys the raw material (the milk) from the peasants and, having reprocessed it into butter, sells it to consumers, but comes into an antagonistic relationship both with them and with the peasants. The co-operative is guided by the will of the workers and it is the interests of their labour which it defends – not the labour interests of the peasant nor the interests of the urban consumer.

Let us then suppose that our workers' partnership is itself forced to close down and that the butter manufacturing factory is bought up by a consumers' association which has been formed in the town and which wishes to supply its members with cheap, good quality butter.

The Processes and the Concept of Vertical Concentration

Scenario C depicts the pattern of economic relationships in this new case of co-operative organization. The managerial power passes entirely out of the hands of the butter producers and is exercised from outside. The production of butter is organized on principles close to those of capitalism and is in an antagonistic relationship both with the peasants and with the workers.

Scenario D assumes the case where the factory passes into the hands of a private entrepreneur – let us say, one of the foremen who has enriched himself – who then organizes the enterprise solely according to his own will and is in an antagonistic relationship with all the other actors depicted in our scenarios.

Thus our factory has remained intact and has gone on working and has, from a technical point of view, remained just the same during all of the four phases of its existence. Nevertheless, in each of these four cases, different relationships arise between the actors depicted in the scenarios, even though they themselves have remained unchanged; and the economic nature of each of the four permutations differs greatly from that of the others. The interests being defended are also different. A peasant butter manufacturing association aims to sell milk at the highest possible price and to pay as few workers as possible. A workers' partnership (*artel'*) receives a payment for its work which depends on the difference between a cheap purchase and a sale at a higher price. A consumers' society aims to obtain butter at the lowest possible price and therefore, so far as it can, it pushes down the payment for milk and for labour. This is identical to the aim of the capitalist who seeks, besides that, to obtain high selling prices for butter.

Thus it is that different kinds of co-operatives differ both in their nature and in their interests. Those who support the unity of the co-operative movement envisage the merger of the first three types – or at least of the first and third types – into a single enterprise. Is this possible?

The difference between these interests is so great that the very numerous attempts to achieve organizational mergers between different kinds of co-operatives with different class affiliations have usually resulted in failure and have raised questions as to whether an integrated co-operative movement is possible.

In one case, in a province on the middle Volga, an association of co-operatives emerged whose members in the south produced wheat and bought felt boots, while those in the north produced felt boots and bought wheat; and they conceived the idea of making an exchange between the co-operatives for the purpose of mutual advantage. Three years after this arrangement had begun, I had occasion to meet the leaders of the Association and to ask how this scheme was working out. 'Things are going well,' so I was told. 'We sell our felt boots in Moscow and buy our felt material in Kazan. We drive the wheat from the southern areas to Moscow and we make purchases for the northerners in Vyatka.' 'And what about exchanges between the co-operatives?' The person to whom I was talking merely waved his hand. It was quite clear that it was the actually existing combination which was the most advantageous.

I think that the case just described can to some extent provide an answer to the question whether an integrated co-operative production is possible.

It is possible for peasant and urban co-operatives not to be mutually hostile, to trade on a reciprocal basis and even to become unified at congresses and in organizations based on some common idea or financial principle. However, the interests that they defend are so much in conflict that they cannot be integrated within one single organization, for its will to exist is inevitably undermined by the inner contradictions between conflicting interests.

Such are the general considerations that compel us to acknowledge that agricultural co-operatives represent an economic phenomenon which resembles other types of co-operative only in an outward and formal sense, but whose nature differs from them profoundly and therefore needs to be made the subject of a separate study.

It must be emphasized that everything set out above was discussed from the point of view of the national economy of a capitalist society.

Having outlined our categories, we now come to the central and most important set of questions of the present time:

1. What are the internal changes which will be required in the processes of the vertical concentration of agriculture, particularly in its co-operative forms, when the regime of a capitalist society is replaced by a transitional system of state capitalism and, after that, by a regime of socialist organization of production?
2. Do we, in our present-day organizational work in relation to the peasant economy, need methods of vertical concentration and, if so, in what forms?

It is not very difficult to answer the second of these questions.

Inasmuch as the gaining of organizational control over the processes of agricultural production is possible only by replacing the diffuse peasant economy by forms of concentrated production, we must in every way develop those processes of rural life which lead towards such concentration.

The path of horizontal concentration, with which we usually

associated the notion of large-scale forms of production in agriculture in a country consisting of small peasant households, was thought of, from an historical point of view, in terms of the spontaneous differentiation of peasant households. Among these, the poorest would turn into the proletarian workforce, while the middle peasants would fall away, production becoming concentrated among the prosperous groups who organize it on capitalist principles and recruit hired labour. This process was bound to lead to the gradual creation of large-scale and technically quite well-organized enterprises which, at the time of formation of a socialist economy, could be nationalized and turned into grain and meat factories.

But it can easily be understood that in the context of Soviet rural policy with its Land Code of 1924 and its general regime of land nationalization, this path is totally out of the question. The proletarianization of the peasantry cannot in any circumstances be a part of Soviet policy. During the course of the revolution we were, indeed, not only unable to concentrate scattered areas of land into large-scale production units, but were obliged to break up a considerable part of the old, large-scale estates.[2] It follows from this that the only form of horizontal concentration which can at the present time occur and actually be achieved is the concentration of peasant land-holdings into large-scale production units in the form of agricultural collectives of various kinds, in the form of agricultural communes, partnerships and associations for joint cultivation of the land – since they are, of course, based on peasant land and not on the old estates.

This process, as we shall see below, is occurring on a considerable scale; but it is not taking place, nor can it take place, on the scale needed for an overall policy aimed at the concentration of agricultural production. Therefore, the most important means of achieving concentration of peasant households has to be one of vertical concentration. It must take co-operative forms, since only in these forms will it be organically linked with agricultural production and capable of acquiring the necessary depth. In other words, the only path which is possible under our conditions for introducing into the peasant economy elements of a large-scale economy, of industrialization and of state planning, is the path of *co-operative collectivization*, the gradual and consecutive separation of particular sectors of specialization from individual households and their organization as public enterprise.

When understood in the way just set out, agricultural co-operation becomes practically the only method of bringing our agriculture into the system that exists in the USSR. And this, at the present time, represents our basic task.

Agricultural co-operatives arose in our country long before the revolution of 1917. They existed, and exist, in a number of capitalist countries. However, in our country before the revolution, as well as in all other capitalist countries, they represented nothing more than an adaptation by small-scale commodity producers to the conditions of capitalist society and a weapon in the struggle for existence. They did not represent a new social system, nor could they do so; and in this respect all the dreams of the many ideologues of co-operation were utopian.

But this state of affairs is radically changing, since the system of agricultural co-operation, with its social capital, its high degree of concentration of production and the planned nature of its work, is appearing not in the context of capitalist society but in the conditions of a socialist society or, at least, of the state capitalism which exists in our country. In this case, precisely because of its high degree of vertical concentration and centralized co-operative system, it becomes linked through its centres with the directive agencies of the state economy and – from being a simple weapon created by small-scale commodity producers in their struggle for existence in a capitalist society – it becomes converted into one of the main components of the socialist system of production. In other words, from being a technical implement of a social group or even of a class, it is being converted into one of the foundations of the economic structure of the new society.

This process of the transformation of the inner social and economic content of the co-operative movement, with the replacement of the political domination of capitalism by the power of the working masses, was highlighted in a particularly vivid way by Vladimir Lenin in the articles *About Co-operation*, which he wrote shortly before his death. After noting the importance of co-operation, in the sense explained above, within the system of state capitalism, he foresaw the possibility of the further development of this transitional form and ended his argument by pointing out that 'a system of civilized co-operators, based on public ownership of the means of production and on the victory of the proletariat over the bourgeoisie, is socialism.'

This interpretation of the national economic importance of agricultural co-operation is, in effect, what predetermines the main thrust of our agricultural policy. Given the nationalization of the land and the political domination of the working masses, this economic system, introduced into the system of the planned state economy by means of co-operative alliances and co-operative centres, can be regarded as being identical to the socialist organization of agriculture.

Such was the origin of the new forms of agriculture, built on the principle of vertical concentration. In its present condition, the co-operative movement in different regions is at different stages of a gradual development. Whereas in some provinces of the USSR we see only the first embryonic signs of buyers' and sellers' co-operatives, places such as the renowned district of Shungen or the districts of Velikiye Soli, Burtsevo and Kurovo in the Moscow province, provide us with many examples of co-operative concentration penetrating into the very core of agricultural production and marketing. By closely following how they developed, we can trace to a certain extent the outlines of the new organizational forms of the agriculture of the future.

NOTES

1. Reference can in any case be made to the excellent new book by V. S. Bernshtein-Kogan, *Vvedenie v ekonomiku promyshlennosti* [Introduction to the Economics of Industry], Moscow, 1926.
2. Editor's note: for contemporary discussion see V. P. Danilov, *Rural Russia under the New Regime*, London, 1987; O. Figes, *Russia's Peasant War*, Oxford, 1989. Also T. Shanin, *The Awkward Class*, Oxford, 1972, Part III; G. T. Robinson, *Rural Russia under the Old Regime*, New York, 1966.

2
The Theory of Differential Optima and Co-operatives in the Peasant Economy

The broad national economic perspectives which we outlined in the last chapter, and the perspectives which we are outlining for the development of agricultural co-operatives, cannot be totally understood or expressed in concrete terms until we become familiar with the peasant household itself, which constitutes the foundation for the building of co-operatives and which is, in its present-day form, the raw material for all measures taken to organize our agriculture.

Despite the fact that over the past fifty years a great deal has been said and written about the peasant household, especially in Russia, we have, properly speaking, only very recently begun gradually to discern, from amid the torrent of general ideas and polemics, certain generally accepted propositions. These have been arrived at by empirical means and repeatedly tested and confirmed, and, as they accumulate, promise to provide us with an objective theory of the peasant household.

We are in any case already in a position to establish a number of propositions, which are at least sufficient to enable us to express our co-operative theory in concrete terms and to establish its social orientation.

The peasant households, taken as a whole as some kind of social stratum, represent a complex phenomenon which is extremely heterogeneous in its make-up. Already by the end of the last century, thanks to the fortunate endeavours of Shlikevich, the statisticians of the *Zemstvo* [regional authorities], when processing statistical material based on the censuses of farmsteads, had begun to use the method of classifying households according to the scales of land used, the area which they had under cultivation and the numbers of their cattle.

It transpired from this system of classification that peasant households varied enormously in size and included those whose area was five to six times greater than that of their neighbours.

Because they were not well versed in studying the organizational foundations of the peasant household, many of the *Zemstvo* statisticians did not pay due attention to the fact that under a communal system of land tenure the land area of households was determined by the number of people, workers and mouths to be fed.[1] Nor did they pay due attention to the fact that precisely this stratification had been revealed by classifications of peasant households described in the censuses of the eighteenth century. They then hastened to identify this stratification of the peasant masses – which at that time had been established only in a *static* form – as a *dynamic* process of horizontal concentration in Russian agriculture. From this point of view, the system of classification according to the area under cultivation revealed the social differentiation between peasant households. It also identified the peasants who sowed on a large scale as constituting the embryo of rural capitalism, and those who sowed on a small scale as the incipient proletarian workforce of the countryside.

This point of view, which seemed extremely plausible from a superficial observation of the countryside, persisted stubbornly and for a long time in economic literature, and it led economic thought on to an unproductive path of research. Only in recent years has it been demonstrated in the writings of N. Chernenkov, G. Baskin, B. Kushchenko and other statisticians that the system of classification according to the area under cultivation cannot, of itself, serve as a tool for revealing the social differentiation of the countryside; and that it largely reflects the demographic process of an increase of families, who obtain for themselves a corresponding increase in land, through communal redistributions or through leasing.

The fact that the system of classification on the basis of area under cultivation was ill-suited for studying the social structure of the countryside was noted also by a number of Marxist economists. Among these we find the sharpest critique in the work of Ya. Yakovlev which was concerned with the grain and fodder balance and was published in the USSR by the Central Statistical Board (TsSU).

The conviction gradually arose, even earlier, that the social structure of the countryside and the processes by which capitalism develops within it have to be studied not through systems of indirect classification on the basis of the area under cultivation, but through the direct study of capitalist relationships in the countryside.

The first to embark on this path was V. Groman who, when dealing with statistical censuses for the Penza province, classified households in relation to the hiring of labour. Similar classifications were used by N. Makarov when calculating work norms in relation to land tenure. Finally, in the summer of 1925 in order to clarify the nature of these relationships in the contemporary countryside, a number of field studies were undertaken by our Research Institute for Agricultural Economics. The material collected on these field studies enables us to identify six basic social types of peasant household:

1. The classical *kulak* household which may at times conduct its agricultural work without hired labour, but which derives the bulk of its income from the trade turnover, from credit based on usury and, in particular cases, from the hiring out of stock and other implements of production to poor households on conditions amounting to enslavement. Here the sources of capitalist income are either the trading profit or superprofit, or the income from the circulation of capital in enterprises belonging to others.

   The development of households of this type has varied in different regions and in different epochs. In general, such households are few in number; but in terms of their influence they represent a major force in the countryside.

2. Households which do not engage either in usury or in trade but which have to be classified as semi-capitalist because in their agricultural or extra-agricultural work they constantly employ hired labour on a large scale, usually in addition to their own labour, in order to obtain an entrepreneurial income from such employment.

   Households of this type are particularly developed in those regions which have a large number of land-holdings and which specialize in producing commodities for export (such as tobacco, wheat, the produce of market gardens, and so on). In these regions, such households are more numerous than those of the first type, but their social influence is always less and they themselves become the victims of exploitation by the *kulak* households.

3. Households which neither hire labour nor engage in other forms of capitalist exploitation. They run their households through the labour of their large families, they are reasonably well provided with the implements of production and they therefore expand the volume of their economic activity. Sometimes they are not inferior to households of the second type, especially when they use complex machinery and a tractor.

   Households of this type are especially common under the

communal system among the old, complex, undivided, patriarchal families with married sons. They comprise a large proportion of the households among the group which cultivates relatively large units. Such households will sometimes hire labour on a day-to-day basis in order to help the family at a busy period or to mow the hay or cart the manure. Such households are usually exposed to capitalist exploitation only as the result of market relationships, or when some of their members engage in off-farm labour.

4. Households which do not employ hired labour and which do not hire out their own labour to other households, but which, owing to the small size of their families or to the shortage of manpower or of the means of production or of land, are unable to develop into economically robust households of the third type. Households of this type are the commonest type of self-employed family peasant farms [*trudovye khozyaistva*] in our countryside. Apart from the customary kinds of market exploitation, they may be exploited by households of the first group, who loan them stock or cattle for productive purposes or grant them credit at commercial interest rates.

5. Households that, because of shortcomings in the use of land or shortages of the means of production or for some other reason, hire part of their manpower to households of the second group or to other employers. Despite the alienation of a part of their manpower, these households nevertheless continue to run a fully-fledged agricultural enterprise with a developed commodity sector.

    This type of household, if one excludes those whose members earn money outside agriculture, develops in a manner parallel to that of households of the second type and is very often the victim of all the kinds of exploitation enumerated above.

6. Proletarian households whose main income is derived from the sale of their manpower, that is, from their wages. These households nevertheless have their own farming activities, usually on a very small scale and nearly always for their own consumption.

It should be noted that when we drew up the above classification, we took no account of the impact of work inside and outside their villages and households. The reason is that in those regions where such economic activities are developed, this form of employment of individual members of families is equally common to all our six types of peasant household.

It is much to be regretted that neither our present nor still less our pre-war statistics, nor the statistics of the West, provide adequate data for measuring in quantitative terms the importance within the countryside of the six types of household listed above. The only available data are those relating to the percentage of households employing hired workers. In this respect, Table 1 provides us with some very revealing figures.

Unfortunately, we have no large-scale data as to the loaning of stock or of cattle for productive purposes, still less do we have any large-scale data on usury. Yet, according to the observations of the Volokolamsk field study of 1925, it was precisely these forms of capitalist exploitation which had become most widespread in the rural areas of the north. Thus, for example, among the households studied in this expedition, 27.3 per cent used work-stock belonging to others.

When completing our account of the different social types of peasant household, we also have to note the very important fact that none of the types which we have listed above is very often encountered *en masse* in a pure form. More often than not, households which employ hired labour will themselves, in turn, hire out their own labour; households which accept credit on terms amounting to usury will, on exactly similar terms, loan out their seed-drills or wooden ploughs.

This situation has become so confused that when preparing the material collected by the Volokolamsk expedition of 1925, its leaders (Ya. Anisimov, I. Varemenichev and K. Naumov), proceeding from the ideas of L. Kritsman, suggested the following method – which

Table 1 *Percentage of peasant households employing workers for fixed periods according to figures of the Central Statistical Agency (TsSU) for 1924*

	Province	%		Province	%
1.	Far North	2.0	11.	Lower Volga	0.8
2.	Northern	1.6	12.	Crimea	0.6
3.	North-West	1.6	13.	Soviet Caucasus	1.2
4.	Western	1.0	14.	Kirghizia	1.6
5.	Moscow, industrial	1.9	15.	Siberia	2.8
6.	Central Agricultural	0.6	16.	Far East	1.6
7.	Volga-Kama	0.6	17.	Turkestan	1.6
8.	Urals	1.6	18.	Byelorussia	1.6
9.	Bashkiria	1.0	19.	Transcaucasus	1.6
10.	Volga	0.8	20.	Ukraine	0.6

although very tentative nevertheless produces interesting results – for depicting the social nature of each peasant household in quantitative terms.

For the social description of peasant households they singled out two types of characteristic: that of *hire* and that of *the use of capital belonging to others*. They measured the capitalist elements in the structure of the household on the basis of the percentage of hired labour in relation to the total labour used within the household; and on the basis of the percentage of capital loaned outside, either in the form of money or of the hiring out of the means of production (stock, etc.). These percentages were added together in order to arrive at a coefficient of the degree to which the households were capitalistic.

In order to measure the degree to which the households were proletarianized, a reverse calculation was made. The resulting series of coefficients demonstrated how intensely the elements of capitalist exploitation were developing in some of the households in the flax-producing region.[2]

Having described the social make-up of our countryside through identifying the six types of household listed above, we can now turn to the basic question as to the importance of agricultural co-operatives for each of these types, and as to the future roles of each of the types identified in the pursuit of co-operative work.

It is absolutely obvious that the first type as such will not only be unable to bring its specific characteristics into co-operative work, but is sharply antagonistic to such work. The whole purpose of co-operatives involved in credit, marketing and the handling of machinery is to deprive this type of household of its basic functions. It is no accident that credit co-operatives in Germany developed and became consolidated as the direct result of their confrontation with usurers. In other words, households of the first type remain, and are bound to remain, outside the ambit of the co-operative movement.

In just the same way, the sixth, most proletarianized, type of household will also remain outside the ambit of the movement. It will do so not because it is antagonistic to co-operatives but because households of this type do not have the possessions which are needed for participation in co-operatives. Their tiny households – which, moreover, only produce for their own consumption – provide nothing for marketing co-operatives; the tiny area of land under cultivation makes the use of machinery unnecessary; and the volume of their purchases and credit will not make it worth while to hold shares in a co-operative. For this reason their active participation in co-operatives is possible only if they move up from the sixth type to the fifth type, which is sometimes possible if they are given co-operative credit.

All the remaining groups, that is the overwhelming majority of households in our rural areas, have possessions which, even though diverse, are sufficient for co-operative activities.

It is true that, with regard to the use of complex machinery, and to some extent also with regard to the purchase of the means of production and even the primary reprocessing of agricultural raw material, households of the second and third types are, because of their size, more able to do without co-operatives than are households of the fourth and fifth types. It should be remembered, on the other hand, that participation in co-operative activities, both with regard to the buying of shares and with regard to work with co-operatives, requires a certain minimum level of material well-being and a certain economic stability, and that this slows down the entry into co-operatives of the weakest households of the fourth and fifth types. For these reasons, we must expect an equal degree of participation in co-operatives by both these types of peasant households.

If one also remembers the greater entrepreneurial flair of peasant proprietors of the second and, to some extent, of the third types, then from a *dynamic* point of view one should even expect them to join in co-operatives earlier than the small households. This is to some degree confirmed by empirical findings, in so far as this is made possible by contemporary investigations of the social structure of the countryside. Thus, for example, on the basis of recently published works by I. Gritsenko, A. Minin and others, we can see the following percentage of mass peasant participation in the co-operative movement, subdivided into groups according to the size of household (shown by the dynamic studies of 1924).

It can be seen from Table 2 that co-operatives are being joined by the households which sow on the largest scale and by households with an average number of cows. Remembering that the number of cows is a better indicator of well-being than the land area under cultivation, it can be seen that these figures confirm our presuppositions.

Bearing in mind, however, the generally insignificant number of households which sow on a large scale and the overwhelming numerical preponderance of households of the middle groups, we have to recognize that among co-operatives, and indeed among the general mass of households, it is the middle strata which predominate.

Of course, the quantitative classifications which we have quoted, especially those relating to the land area under cultivation, are, as we have already observed, far from adequate for the purpose of giving a social-economic description of households involved in co-operation. However, even they lead us to suppose that, while there is an overwhelming preponderance in the countryside of the middle strata, the other strata also take part in co-operatives.

The social foundation of agricultural co-operatives rests on two main groups of peasant household:

1. Households based on the employment, to a greater or lesser degree, of hired labour, for the purpose of obtaining an entrepreneurial income from the exploitation of this labour. In their pure form – that is, in reference to households based exclusively on hired labour – we shall henceforward term them *capitalist market-oriented farms [kapitalisticheskiye tovarnye khozyaistva]* (type 2).

2. Households which derive the overwhelming proportion of their output from the labour of the householder's family without employment of hired labour for entrepreneurial purposes; or, in a pure form, households which do not employ hired labour at all or else offer their own labour for hire. These we shall henceforward refer to as market-oriented family households [*semyeinye-tovarnye khozyaistva*], or simply as market-oriented peasant households [*tovarnye krest'yanskiye khozyaistva*] (types 3, 4 and 5).[3]

There exist in between these two a good many intermediate types: and, in its pure form, the capitalist market-oriented farm is at present encountered in Russian peasant conditions only very rarely. However, both of these types represent the main tendencies in the economic organization of the peasant homestead.

It would be a very great mistake to confuse these two types of enterprise. Each of them has its own organizational peculiarities and they not infrequently differ in their economic behaviour.

Readers who want to familiarize themselves more closely with the distinctive organizational features of these two types can get the necessary information from a number of special investigations.[4] In this book we shall therefore confine ourselves to the comparisons needed for the purpose of our subsequent explanations.

The main elements in the economic organization of the capitalist market-oriented farm are: 1. its gross income; 2. the material costs of production expended in kind or in money or in deductions for amortization; and 3. the wages actually paid to the wage-earners. 4. Gross income less material costs of production and of wages, which are covered by the capital advanced, produce 5. profit which is the sole entrepreneurial purpose of the farm. The profit is not connected with the labouring activity of the proprietor's family and it depends, other things being equal, solely on the amount of the capital advanced to the farm.

Table 2 Social composition of peasant household members of co-operatives in the USSR (1924).

Groups classified according to the area under cultivation per household

Desyatiny of land sown	0.0	0.1–1.0	1.1–2.0	2.1–3.0	3.1–4.0	4.1–6.0	6.1–8.0	8.1–10.0
Membership of agricultural co-operatives per hundred households of given group	2.5%	2.8%	3.2%	3.9%	4.0%	4.8%	5.7%	7.1%

Desyatiny of land sown	10.1–13.0	13.1–16.0	16.1–19.0	19.0–22.0	22.1–25.0	25.1–30.0	30.1–40.0	40.1 and over
Membership of agricultural co-operatives per hundred households of given group	9.1%	9.6%	10.3%	10.7%	11.0%	11.0%	16.0%	17.5%

Groups classified according to the number of cows per household

Number of cows	0	1	2	3	4 and over
Membership of agricultural co-operatives per hundred households of given group	3.8%	3.9%	6.6%	5.7%	6.2%

Groups classified according to the area under cultivation per household

Desyatiny of land sown	0.0	0.1–1.0	1.1–2.0	2.1–3.0	3.1–4.0	4.1–6.0	6.1–8.0	8.1–10.0
Proportion of households in co-operative system which have this area under cultivation	3.2%	9.5%	16.8%	17.3%	13.0%	17.7%	9.0%	5.1%

Desyatiny of land sown	10.1–13.0	13.1–16.0	16.1–19.0	19.1–22.0	22.1–25.0	25.1–30.0	30.1–40.0	40.1 and over
Proportion of households in co-operative system which have this area under cultivation	4.1%	1.8%	0.9%	0.5%	0.3%	0.3%	0.3%	0.2%

Groups classified according to the number of cows per household

Number of cows	0	1	2	3	4 and over
Proportion of households in co-operative system which have the number of cows shown	16.1%	55.7%	22.3%	4.1%	1.8%

Note: 1 Desyatina = 2.7 acres.

As a result of this, the size of a capitalist farm can in theory be expanded without limit by the hiring of more and more workers, while the degree to which intensive economic methods are used and the choice of the crops and specialized sectors which comprise the organizational plan and the way they are combined, are all totally determined by the influence which they may have on the size of profit (or, to be even more precise, on the rent, that is, on net income less the usual rates of interest on capital).

A market-oriented peasant family household builds its organization in a different way and in its case the main elements are: 1. the same gross income as above; 2. the same material costs, but instead of wages actually paid we have here the labour in kind provided by its family.

It is therefore impossible for us to identify any genuinely distinct or physically perceptible net income in a family household. The only reality which exists is that of the gross income. If one subtracts from this gross income the sum of the expenditure in real terms which is incurred in order to reproduce the capital for the purpose of the following year's output, then there remains available to the household's family a sum which can be set aside either for personal consumption or for capital accumulation. This sum has been defined by Professor E. Laur as the payment for the labour of the peasant family (*Arbeitverdienst*). It is, indeed, the aim of the family household to earn this sum, together with the opportunity of completely reproducing its real capital every year. In this case, to use the customary language of political economy, the interests of the peasant as an entrepreneur and his interests as a worker are inseparably and indivisibly merged into a single whole.

In view of the fact that in market-oriented family households work is performed by the family itself, the total amount of labour expended and, therefore, the overall volume of economic activity – given an adequate supply of the means of production – is determined (or at least restricted) by the size of the family, that is, by the number of its members who can work. It must be reckoned, however, that the question in this case concerns the maximum possible volume of economic activity to which the household aspires. The actual volume of economic activity usually falls short of the maximum in view of the shortage of capital and of the means of production which is usual in our peasant households.[5]

These, then, are the organizational patterns of the two types of enterprise. Where land is relatively abundant and where family households are able to employ their manpower more or less to the full, these two types of household differ little in their economic behaviour because a high remuneration per unit of labour coincides with a high net income.[6]

However, the situation begins to change when we switch our attention to the areas of agrarian overpopulation, that is, to areas where the historically conditioned size of the peasant population significantly exceeds the manpower needed to cultivate the land at the level of agricultural intensity which yields an optimal net income.

A capitalist farm, in such conditions, would simply dismiss the workers which it did not need. But a family household which is unable to apply its labour to its small plot of land is in a different position; and since the peasant cannot dismiss himself from his own household he finds himself in an enforced state of partial unemployment. If, because of the circumstances, there is no way out of the situation through wage earnings, through crafts or through land leasing, then the position of the household becomes doubly distressing: its manpower is reduced to enforced idleness while the family's consumer budget is sharply curtailed. This prompts the householder to search for some way out of the situation which has arisen.

It is in these conditions that the difference between the family household and the capitalist farm begins to make itself felt: because the interest in the largest possible gross income and the interest in obtaining the largest possible *annual* remuneration of labour begin to outweigh the interest in the highest possible remuneration per unit of this labour. Unsatisfied needs, combined with a surplus of labour which is unable to find new means of employment, begin to weigh down on the household, compelling it constantly to seek out new ways of applying its labour on the same area of land, at the cost of a sharp drop in the remuneration per unit of labour.

It thus quite often happens that work which is deemed by a capitalist farm to be unprofitable is profitable in a market-oriented family household, and vice versa. For example, let us try to explain why, during the decades immediately before the war, the peasant households in the Tver and Smolensk provinces very willingly and energetically expanded the areas under flax cultivation although capitalist landlords almost entirely refrained from sowing flax.

Table 3 gives us a computation of income and expenditure for each *desyatina* [2.7 acres] of land under flax cultivation with a similar computation for land cultivated by oats.

For a household which is based not on its own members' work but on hired manpower, the sowing of oats is undoubtedly more profitable than the sowing of flax, because the profit from oats is twice as high per unit of land. And if we express it as a percentage of

Table 3 *Approximate production costs and income yield per hectare of land under flax and oats (on one's own land)*

Expenditure	Oats	Flax
Seed, horses, etc.	15 roubles	15 roubles
Human Labour	20 roubles	80 roubles
	(20 working days)	(80 working days)
Total expenditure	35 roubles	95 roubles
Gross income	45 roubles	100 roubles
Net income	10 roubles	5 roubles

the entrepreneur's capital outlay (as a percentage of expenditure), then the advantage of sowing oats is even more substantial, since the net profit in relation to the outlay of turnover capital will be 28.6 per cent for the sowing of oats while for the sowing of flax it will be no more than 5.3 per cent.

It is therefore quite understandable why capitalist farming avoided the cultivation of flax; and, according to the 1916 agricultural census, out of the entire area of land under flax cultivation only 3.1 per cent was attributable to the estates of large landowners.

Yet for a peasant economy in a state of agrarian overpopulation, the sowing of flax may turn out to be preferable to the sowing of oats, since it provides an opportunity for the fullest use to be made of the family's manpower and enables the family to obtain from that same *desyatina* [2.7 acres] a remuneration of 85 roubles for its labour, instead of the 30 roubles obtained from oats.

It is true that by applying 80 working days of his labour to a *desyatina* of land under flax cultivation, a peasant can, for each working day, obtain a remuneration equal to the value of the finished product, that is, 1.06 roubles, whereas a working day expended on the sowing of oats is remunerated at the rate of 1½ roubles.

There is no doubt that if a peasant were able to achieve a five-fold expansion of his sowings, and if he were able to apply all his labour to the sowing of oats, then he would not stand to gain from the cultivation of flax. However, should the tiny peasant plots in the areas of agrarian overpopulation sow nothing but oats, they would condemn themselves to enforced unemployment for the greater part of the year. Anyone who has observed a peasant household in areas of agrarian overpopulation in Russia can clearly see that the peasant plot, given the present three-field farming system, is not only incapable of feeding the household's family with its harvest produce, but is unable to utilize even half of its manpower.

It has been proved by a number of statistical investigations that as a rule, a peasant family is able – on its own land and given the three-field system – to utilize between one fifth and one quarter of its available working time; and it is, of course, far from being able to earn enough income from the land to cover all the expenditure which is essential for its livelihood. Therefore a peasant household, when it has unused manpower and far from adequate resources for a livelihood, naturally seeks to find an outlet for its unused labour in order to increase, one way or another, its annual earnings.

When seeking an outlet for its labour, such a household will often accept a very low remuneration and is even prepared to undertake economic activities of a kind which – according to any calculation of the family's labour in terms of ordinary wage rates – not only bring no profit, but appear to produce an undoubted loss.[7]

Nevertheless, the peasants do undertake such work: they pay loss-making land-rents, they engage in unprofitable cottage industries, they sow crops on their fields which require a great deal of labour and yield a high gross income per unit of land, but which provide low remuneration for every working day expended. It is self-evident that all this is done in those cases where there is no other, more profitable, outlet close to hand for the employment of labour.

No one can feel enthusiastic about such a state of affairs. Agrarian overpopulation and the things that go with it are one of the most terrible scourges of our national economy. The struggle against it, as well as the struggle against the decaying three-field system based on extensive methods which exist in these areas, is an urgent task of our economic policy.

However, it must be noted that the misfortune in this case is the fact of agrarian overpopulation and all that goes with it – and in no way the ability of the peasant economy to adapt itself to this calamity. The peasant economy's flexibility and its ability to adapt itself in the face of the most difficult conditions of existence should be regarded as a very great virtue of its economic organization.

This capacity for resistance saves the peasant economy not only in cases of agrarian overpopulation but also in times of violent market fluctuations which totally ruin enterprises organized on capitalist principles. Let us suppose that a capitalist entrepreneur contemplates sowing five hectares [12.4 acres] of oats on leased land and wishes to ascertain whether this undertaking is advantageous.

His economic calculations will work out roughly as shown in Table 4. The fact of a net profit of 40 roubles, amounting to about 12 per cent of the capital outlay (320 roubles), will make the operation profitable and the entrepreneur will try to undertake it, since

Table 4

Expenditure

Leasing of 5 hectares multiplied by 16 roubles	80 roubles
Seed	50 roubles
Gathering in of the harvest. 100 working days multiplied by 1 rouble 20 kopecks	120 roubles
Work with horses	30 roubles
Amortization and other overhead expenses	40 roubles
Total expenditure	320 roubles

Receipts

88 quintals of oats at 375 kopecks each	330 roubles
48 quintals of straw at 62.5 kopecks each	30 roubles
Total receipts	360 roubles
Net income	40 roubles

investments of the capital in interest-bearing securities or by deposit in a bank, will yield a significantly lower profit (5 to 7 per cent).

But the situation will become entirely different if the price of oats is not 375 but, let us suppose, 312.5 kopecks per quintal. In that case, the calculation will appear as follows:

Income from the sowing	305 roubles
Expenditure	320 roubles
Loss	15 roubles

An undoubted loss will make the undertaking unprofitable for the entrepreneur and he will abandon sowing since he cannot undertake it without loss.

The peasant family – whose basic aim is not to obtain a rate of interest on comparatively minute capital, but to obtain remuneration for a year's labour – will make its calculations on an entirely different basis.

1. The peasant family will invest 100 working days in the land sown.
2. In addition it will spend 200 roubles in rent, in payment for seed and for the work of horses and on other expenses.
3. If the price of oats is 375 kopecks per quintal it will earn 360 roubles.

4. After subtracting from income the material costs which it has incurred, the family will earn, as a result of its labour, 160 roubles.
5. These 160 roubles represent the remuneration for the 100 working days expended by the family on the sowing of oats; consequently, the remuneration for one working day invested in this activity has been 1 rouble 60 kopecks.

This then is the remuneration for his working day which the peasant would compare with other possible rates of remuneration for his activities. And he will regard the sowing of oats as profitable once he has become convinced that no other activity will remunerate his labour at a rate above 1 rouble 60 kopecks.

For the peasant, this and this alone is the yardstick for ascertaining the profitability of alternative activities. If, let us suppose, an agricultural household pays for his work at the rate of 80 kopecks and work in crafts pays at the rate of 1 rouble, then he will 'have no time' to work in agriculture. The peasant uses the same yardstick when comparing the profitability of crops in his own household.

It need hardly be said that such a method of ascertaining profitability may lead to conclusions which are the exact opposite of the reasonings of the capitalist entrepreneur. Thus, for example, we have demonstrated that if the price of oats is 50 kopecks per quintal, the entrepreneur stands to gain nothing by sowing oats on leased land. But let us see whether a peasant can sow them. For this purpose we shall repeat the calculation already quoted:

1. The peasant family invests 100 working days;
2. In addition it will incur costs of 200 roubles for rent, seeds, etc.;
3. If the price of oats is 3.125 roubles per quintal, it will receive 305 roubles;
4. After subtracting from income the material costs incurred, the family will earn 105 roubles;
5. 105 roubles represent the remuneration for 100 working days; consequently one working day expended on the sowing of oats will be remunerated at the rate of 1.05 roubles.

But is the remuneration of labour at the rate of 1.05 roubles for one working day possible and acceptable for a peasant worker? It is not, if other earnings at a higher rate are open to him. But the answer is undoubtedly yes if no other, more profitable, earnings exist. In other words, whereas in an enterprise organized on capitalist principles, there appears a loss which erodes the farm's material capital, in the case of a peasant household we are seeing a drop in the level of consumption, sometimes, it is true, almost down to the level of starvation.

No one, of course, can welcome peasant hunger; but one cannot fail to recognize that in the course of the most ferocious economic struggle for existence, the one who knows how to starve is the one who is best adapted. All that matters is that those who shape economic strategy should make this latter method superfluous. In any case, the peasant family household's capacity for resistance, which we have already noted, and which does not need to manifest itself in extreme forms involving hunger, serves to explain the peasant household's tenacity for life and its astonishing capacity for revitalization during various critical periods of its historical existence.

This is one of the capacities of the market-oriented family farm. We may confidently reckon that the development of agro-technology, the increase of capital investment in the peasant economy, its industrialization and mechanization, a properly organized resettlement of rural populations and, finally, the development of co-operatives, will in time render the manifestation of this capacity superfluous. But the fact must not be overlooked that in the numerous economic crises that still face our co-operatives, the exceptional capacity for resistance on the part of peasant households will more than once make it possible to deflect economic blows away from the co-operative apparatus on to the peasantry – thus rescuing co-operatives from inevitable destruction and paying the production costs of the new national economic system which we described in the first chapter.

These are the attributes of one type of peasant household, upon which agricultural co-operatives base their structure. Another type, which is based on the more or less developed exploitation of hired labour, is obliged, during years of severe crisis, to rely on reserve capital and funds. On the other hand, since its production is on a large scale, it gets the opportunity to utilize the advantages of large-scale farming and, in normal years, to produce great quantities of a homogeneous product at a low cost.

What then is the organizational significance of co-operatives themselves for the households described?

The two types of enterprise described – the capitalist market-oriented farm and the market-oriented peasant household – differ in their methods of economic calculation and therefore in the structure of their crops and specialized sectors and in the extent of their reliance on intensive economic methods. But they do not differ greatly from each other with regard to the technical organization of production itself; and they can therefore be examined together.

The organization of production in any agricultural enterprise will – except in certain special kinds of enterprise – nearly always begin by determining its scale, which is ultimately expressed by establishing a definite area of land for farming.

For large-scale cultivators, the land area was itself the starting-point. For the entrepreneurial capitalist farm, the determining factor is the amount of the capital advanced. For the family peasant household – if it has not been confined within the immovable boundaries of a plot – the determining factor is the existence of both a family and capital.

Only after determining its area is it possible to embark on the organization of the enterprise. Here, the first step is to determine in what directions it is to specialize; that is, in the case of a market-oriented enterprise, to determine the basic commodities which it would be most profitable to manufacture – given the current state of the market and the enterprise's characteristics. Having determined its basic market goals, or, as is usually said, the basket of commodities, the enterprise has to consider them in relation to two other balances: those relating to *fodder* and *consumption* or, more precisely, those ingredients which have to be produced by the enterprise in kind.

Having thereby sketched out a rough organizational *pattern* for the enterprise it is possible to embark on the preliminary organization of the particular piece of land, that is to divide up the land according to its economic purpose: into forest, common pasture, meadow and arable land; and to divide up the latter into fields for grain, flax, fodder, intertilled crops, etc.

When this outline has been drawn up, it is possible to begin organizing the tillage, fixing the seed turnover and composition of the crops, and making estimates with regard to sowing and harvesting. When the tillage requirements have been ascertained, the traction requirements can be calculated: that is, it is decided how many animals and vehicles are needed to work the fields and on what scale they are needed, allowing for the normal care of other land. The cattle, once its composition is determined, is correlated with the amount of fodder which it needs and with the supplies of fodder available to the enterprise – which can in case of need be supplemented by purchases of fodder, either in order to increase the overall number of fodder units within the enterprise, or to ensure that the general fodder supply contains the proper balance between albumen and carbohydrates. Having fixed the composition of the fodder, we can then go on to organize the cattle-rearing. Having completed the organization of the cattle-rearing, we can calculate what manure fertilizer is required, how it is to be distributed and generally organized. After this, it becomes possible to organize specialized sectors: market gardens, orchards and hemp fields.

This completes the organization of the productive sectors and we can arrive at a calculation of the overall expenditure of labour which is required, the distribution of this labour over time and the extent of of its mechanization. After deducting the requisite amount of manpower, we arrive in regard to the family household at a guideline figure which, when compared with the manpower available to the household, enables us to ascertain whether the organizational plan drawn up is, or is not, attainable. The next thing is the organization of the stock and of the technical operations (mainly the primary reprocessing of agricultural products: butter manufacture, the reprocessing of flax, etc.) and the organization of the buildings and management of the farm. This completes the farm's technical organization, expressed in physical terms. It may be verified by a special calculation of the household's turnover of nutritive substances which shows whether the plan drawn up may involve despoiling or impoverishing the land.

After completing the physical and technical organization of the farm, and after making an economic calculation in the form of a financial plan, the farm's organizer proceeds to a final economic calculation in regard to the household, by preparing estimates for the household's output and for its anticipated annual balance-sheet.

This is almost inevitably the way that economic calculations develop in those cases where an organizational plan is drawn up consciously. But if – as happens in the overwhelming majority of peasant households – this plan is evolved, in the manner of the species of the animal kingdom, through a prolonged natural selection of the fittest, then the relationships which we have been examining exist in the economy without anyone being subjectively aware of them. A peasant runs his household in accordance with a definite organizational plan although he is often totally unaware of it, like Molière's Jourdain who spoke in prose for forty years without guessing that he was doing so.

As a result of these and similar arguments concerning the economic and natural conditions for the existence of a household, there arises a correlationship – which differs greatly in different areas and among different social groups – between the specialized sectors of a household and its overall organizational plan.

An example of such a complex organizational plan can be seen in Figure 1, which gives a graphic representation of the economic turnover of a peasant household in Starobelsk, in the Kharkov province.

The Theory of Differential Optima and Co-operatives 43

Figure 1: Economic turnover of an average peasant household in the Starobelsk district of the Kharkov province (1910). Figures refer to roubles

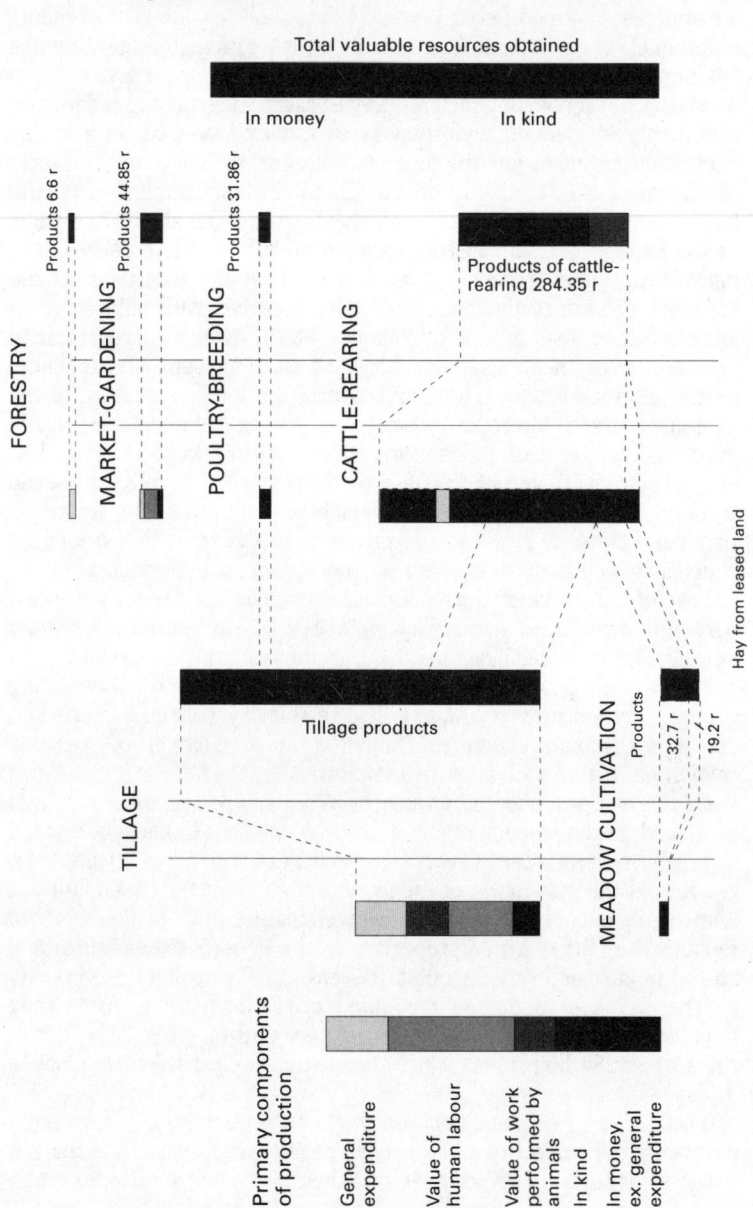

Figure 1 provides an extremely graphic representation of the annual turnover of valuable resources in an average-sized peasant household in the Starobelsk area. Beginning from the left-hand side, we can see, grouped into a vertical column, all the primary elements of production (the expenditure, the outlays of livestock and human labour, the material outlays in money and in kind, which are common to these households). Each of these categories of expenditure is tentatively shaded in; and records of money transactions are also preserved for inclusion among items of receipt. The height of each section of the column is proportional to the sum expended on the household under this section; and the scale for translating the value of the expenditure into figures is given on the right-hand side of the figure. All these expenses are then divided up according to the separate sectors of the household and, together with data of gross incomes, they form groups of columns, each of which corresponds to one sector of the household. Most of these groups are to some extent interconnected. The foundation of the whole structure of the diagram is the field crops category. Its primary elements are made up of various items of expenditure amounting to 306.27 roubles. The above-mentioned sum of valuable resources, after they have passed through the production process, which is denoted on the figure by two vertical lines, yields tillage products of the total sum of 585.63 roubles, which is also depicted by the appropriate column.

Part of the product thus obtained was sold (as shown in black), part of it was placed at the disposal of the peasant householder, and yet another part was again used for production and, as shown by the dotted lines, it passed into the poultry-breeding and cattle-rearing groups. Immediately below the field husbandry columns, there is a group of columns corresponding to the circulation of valuable resources in the cultivation of meadows.

Of the hay produced at a value of 32.70 roubles, a small part was sold; and all the remainder, as shown by the dotted lines, was used for cattle-rearing. Cattle-rearing, which gets its fodder stocks from tillage and the cultivation of meadows, itself involved expenditure in kind on feeding the herdsmen, as well as the cost of the work of looking after the cattle, a proportion of the general expenditure; and all the money expenditure on cattle-rearing. This column represents all the valuable resources expended on cattle-rearing. After they have passed through the production process they yield a product of the sum of 284.35 roubles which means that the cattle-rearing makes a loss.

The group of columns relating to poultry-breeding is constructed in exactly the same way. The groups relating to forestry and market gardens have no connection with the other groups. The extreme right-hand column indicates the sum total of valuable resources obtained as the result of all the production processes. Since all the columns are built to the same scale, the figure makes it possible to analyse not only the organization of every sector of the household, but to compare the relative importance of these sectors and of the interrelationship between them.

We get an entirely different picture from two other cases which represent the correlation between sectors in an exclusively flax-cultivating household in the Volokolamsk district and a market-oriented dairy farm in the Vologda province. In the first case, the household's basket of commodities is based exclusively on field crops; and in the second, almost exclusively on cattle-rearing. There is also a very different structure relating to consumption and to fodder.

By breaking down the elements in these two programmes into their technical components we can see that agricultural production consists of numerous technical processes of differing natures which we can divide into the following categories:

1. Mechanical processes arising from the fact that land extends over space (tilling of the soil, sowing, transportation, gathering of the harvest, driving of cattle, etc.).
2. Biological processes of plant-growing and cattle-rearing (cultivation of plants, milking of cows, fattening of livestock, etc.).
3. Mechanical processes of the primary reprocessing of raw material obtained (threshing, separation of cream from milk, manufacture of butter, scutching of flax [i.e. dressing the flax by beating], etc.).
4. Economic operations linking the household with the outside world (buying and selling, credit relationships, etc.).

In a technical sense each of these operations must have identical purposes, both in a large-scale and, equally, in a small-scale farm. However, some of them are better suited to a large-scale farm, and others to a small-scale farm.

A significant majority of the processes of the first category can be carried out equally well regardless of the scale of the enterprise. A large-scale enterprise has a certain advantage with regard to the use of complex machinery; and a small-scale enterprise with regard to internal transport.

Processes of the second category are considerably better suited to a small-scale farm, since they require great attention and individual care. The only thing that is better suited to a large-scale farm is the process of stock-breeding since the employment of stock-breeders is beyond the capacity of small peasant householders.

All the processes in the third and fourth categories can be considerably better organized in the most large-scale forms.

A detailed analysis of the various tendencies which we have noted towards either the enlargement or the reduction of scale in the different technical sectors of the household leads us to draw conclusions which are of crucial importance for our theory of co-operation.

The most important thing for an agricultural enterprise is not that it should be very large or very small, but that it shoud be of some intermediate, *optimum*, size where the advantages and disadvantages of a small-scale and a large-scale enterprise are balanced against each other. This may now be regarded as proved – by work undertaken by the Research Institute of the Agricultural Economics for the purpose of ascertaining the optimum size of agricultural enterprises, some of whose findings we summarized in Chapter 1. We showed that each system of farming had its own optimum scale of enterprise.

We made all these calculations in relation to agricultural enterprises as a whole. However, further analysis showed that if the organizational plan of the household was broken down into its different sectors, then it would be possible to ascertain for each sector its own peculiar optimum scale. There would be one optimum for the produce of meadow cultivation, another for tillage; and besides that, one optimum for grain crops, one for intertilled crops, another for seed production and yet another for different forms of reprocessing – varying in each case and, as a rule, varying very greatly, and so on.

In other words, the optimum scale for the enterprise as a whole is in no way the optimum for each of its sectors; and in order to get the very best results from applying the notion of an optimum to economic organization in agriculture, we need to forget about the oneness of an agricultural enterprise and to make an organizational breakdown of the organizational plan of an agricultural enterprise into its basic components. We then need to organize each component separately and autonomously on the specific optimum scale which is appropriate to it.

This theory, which we put forward several years ago and which came to be described as the theory of 'differential optima', is one which at first sight seems paradoxical and incapable of being implemented. Nevertheless, if one surveys the practice of co-operative structuring it is not hard to see that it is precisely here that our theory finds its full realization. It can even be said that the theory of differential optima is the basic organizational idea which underlies agricultural co-operation; and that only through co-operation can the theory be put into practice.

Because of the fact that it is technically possible to achieve a breakdown of its organizational plan into its individual components, the peasant household has been able to separate all those mechanical and economic operations whose technical optimum scale was greater than that of the peasant household, from the remaining operations. And it has been able, together with other similar households, to organize these operations on a large scale and indeed on an optimum scale, in a co-operative form.

Nevertheless, those operations whose optimum scale did not exceed that of the peasant household remained totally in the hands of the family household.

Thus, in the first category of technical processes, the use of machinery and engines was allotted to a special machinery association, while in the second category the selection of cattle and regulation of standards for the feeding of cattle were separated off, as also was the reprocessing of milk into butter in the third category. These were organized into appropriate co-operatives. Almost all of the fourth category of economic processes was organized entirely on co-operative principles in the form of purchasing and consumer associations, marketing associations and credit co-operatives. It must at the same time be noted that none of these processes, when organized on co-operative principles, lost their economic ties with the parent economy. And they imparted to the co-operative all those special features of economic organization, as well as that same exceptional capacity for survival, which we discovered in the peasant economy.

Capital within a co-operative plays the same ancillary role as in a family household. The scale of a co-operative enterprise is determined – as in the case of a peasant household – not by the amount of available capital but by the needs of the combined households.

Consumer and purchaser co-operatives cannot have a greater turnover than the purchasing power of their members. The size of a butter-producing factory is determined by the amount of milk available to its members. The credit turnover of a credit association corresponds to the credit turnover of its members, and so on.

The very structure of a co-operative and the profitability or otherwise of what it does are likewise determined not by the quest for a maximum profit on the capital invested in the enterprise, nor by the interests of the co-operative institution itself, but by the incomes from the labour of its members earned through the co-operative, and by the interests of their households.

A co-operative will be extremely useful, therefore, even if it produces absolutely no net profit as an enterprise, but nevertheless increases the incomes of its members. And conversely a co-operative will be harmful if, for example, if produces 10,000 roubles' worth of profits, but if, owing to unskilful management, the peasants suffer a shortfall of 40,000 roubles of income from their labour. The success of co-operatives is measured by the growth in their members' incomes, and not by the profits of the co-operative itself. There is the members' income and nothing else.

The nature of co-operatives, as an integral component part of the family household, becomes particularly visible from a comparative analysis of the capacity for survival of a co-operative *vis-à-vis* a capitalist enterprise. Let us, for example, imagine a co-operative system for the marketing of eggs.

A private trader engaged in the egg trade buys eggs in order to sell them on at a higher price. If, after buying this commodity, the price of eggs on the market substantially falls, then the private trader is compelled to sell them below the price which he paid and makes a loss. But the co-operative is in a different position. It does not engage in any trade, it does not buy a commodity; and for it, the difference between the purchasing and the selling price is of no importance. Co-operation represents the organized marketing of the products of the peasant's labour; and if the price of eggs falls, this means that the peasant households which sell their output through the co-operative will get a lower remuneration for their labour; and, because of the special features of the peasant household, neither the co-operative apparatus nor that of the peasant household need suffer.

Such are the reasons which compel us to acknowledge that co-operation in the villages has no self-contained existence of its own but is a collectively organized extension of family production living the same life as the parent organism.

Professor A. Chuprov once noted in one of his writings that, in relation to agriculture, the idea of co-operation was no less significant than all the most important technical discoveries. And we can indeed acknowledge that the spontaneously evolving method of splitting up organizational plans into individual groups of processes, and of organizing each of them in accordance with their optimal economic and technical parameters, is providing agriculture with a most excellent economic apparatus.

What has just been said serves, in outline, to explain the importance of co-operation for the peasant economy; but it also points to ways of studying this economic phenomenon and of making this into something systematic.

In essence, no system of classification can lay claim to the absolute truth and no classification can be exclusively correct. It can only be the simplest or the most convenient one in relation to the goals which it is meant to serve. We are therefore prepared to accept the validity of any classification which serves the purposes intended by its authors.

When examining the emerging trend of the co-operative movement we must try, first of all, to classify all the diverse things observed into homogeneous categories in order more easily to understand the diversity of types and forms within the phenomenon being studied.

It very naturally follows from our concept of co-operation that this preliminary systematization of the data has to be carried out by reference to those economic processes which are to be organized on co-operative principles. Once we have acknowledged that a co-operative is merely a collective method of organizing the individual components of the organizational plan of a peasant household, it follows that when we build a classification system we must inevitably examine the role of each particular type of co-operative system in this organizational plan.

If in our mind's eye we envisage the organizational plan of an agricultural household, and if we then break it down into its constituent elements and think about which of those elements are most suitable for large-scale organization, this will enable us to determine all conceivable types of co-operation.

It is clear to us that large-scale purchases of the means of production and of articles of everyday use are more advantageous than are small-scale retail acquisitions – and this provides a basis for purchasers' co-operatives. It is equally obvious to us that a large-scale enterprise will obtain credit more easily and more cheaply than a small-scale family household, and that a large-scale organization will be able to sell its products more advantageously than a small-scale household. Hence the bases for credit and marketing co-operatives. The reprocessing of milk into butter, the scutching of flax and the drying of vegetables and fruits are most cheaply and efficiently carried out in factory conditions. Hence there are grounds for organizing co-operative workshops. A large-scale unit is in a better position where the use of complex machinery and stock-breeders is concerned. Hence a further dimension of possible co-operative work, and so on.

By thus analysing the organizational plan of the peasant household and identifying those elements within it which are suitable for co-operation, we can easily draw up a long list of possible forms of co-operation. However, when comparing such a list with real life, we may note that many of the types of co-operation which we have identified analytically do not yet exist in reality. They have not yet been discovered in practice, just as our chemists have not yet discovered many of the elements of Mendeleyev's table of chemical elements. There is no doubt that the co-operative movement will, as it develops and takes root, bring to life more and more of the kinds of co-operation which we have identified.

When classifying particular forms of co-operative into more general categories, we can single out the following headings:

1. *Economic processes: the mechanical processes arising from the space of the land being worked*
    (a) Machinery users' associations.
    (b) Associations for joint ploughing, etc.
    (c) Associations concerned with water and land melioration.
2. *Biological processes*
    (d) Bull pedigree associations.
    (e) Societies for systematic stockbreeding.
    (f) Associations concerned with quality inspection.
    (g) Associations concerned with the selection of breeds.
3. *Mechanical processes of primary reprocessing*
    (h) Threshing associations.
    (i) Butter manufacturing associations.
    (j) Potato-grinding, vegetable-drying and similar associations.
4. *Economic operations linking the household to the outside world*
    (k) Purchasers' co-operatives.
    (l) Marketing associations.
    (m) Credit associations.
    (n) Insurance associations.

This classification, which is derived from the organizational plan of the peasant economy, is somewhat unusual in contemporary co-operative thinking. It is presented in order to help clarify the importance within the peasant household of particular forms of co-operation. But we shall henceforward pursue our study, not in the sequence of the system shown above, but in an order which is more usual and which is more appropriate to the historical sequence in which forms of co-operation actually developed and to the importance of these forms for our co-operative movement.

From an organizational and historical point of view, by far the most important place in our country belongs to the system of co-operative credit; and it is with this that we shall begin our analysis. Apart from that, the oldest and most firmly established type of co-operation is that which relates to purchasing. The next place is occupied by marketing co-operatives and, after that, by co-operatives concerned with reprocessing linked with marketing.

After completing our analysis of these most highly developed and important branches of the movement, which are almost entirely connected with the external relations of the household, we shall then go on to clarify the role and the forms of the co-operative movement in the organization within the household of biological and mechanical processes – in relation to associations concerned with machinery, inspection, stockbreeding and others of a similar kind. Here, we shall devote special attention to those forms of peasant co-operation which are, at the present time, the newest and the least studied.

In accordance with our plan of work we shall attempt first of all to establish in each case the economic nature of the phenomenon which is to be brought within the ambit of co-operation. By analysing its nature we shall establish what organizational problems confront co-operative organizers in each of the cases studied and what are the organizational forms which life has actually put forward in order to solve these problems.

NOTES

1. Editor's note: For a discussion of the Russian peasant land tenure and social organization, see G. T. Robinson, *Rural Russia under the Old Regime*; also T. Shanin, *Russia as a Developing Society*, Newhaven, 1979.
2. Editor's note: The methodology of analysis was fully developed by V. Nemchinov: see *Sobrannye sochineniya*, Moscow, 1970, Vol. 1.
3. Since our concern here is to study co-operatives, we shall not include economies of the natural type in our analysis. The natural economy is alien to co-operatives in the present-day meaning of the word.
4. The organizational foundations of the market-oriented family household are explained in N. P. Makarov, *Organizatsiya khozyaistva* [The Organization of the Economy], Moscow, 1925; and in A. Chayanov, *Organizatsiya krest'yanskogo khozyaistva*, published as *The Theory of the Peasant Economy*, Madison, 1982. When studying the organization of the capitalist commodity economy use may be made of any course on the organization of a large-scale economy.
5. The factors which determine the scale of a family household are extremely complex and we would refer readers who want to familiarize themselves with this question to chapter three of our book referred to in note (4) above.
6. The differences will always be determined by the need of the self-employed peasant family farm [*trudovoye khozyaistvo*] to achieve a more even distribution of labour throughout the months of the year and by the opportunity available to big capitalist enterprises to utilize the advantages

of a large-scale economy (complex machinery, etc.).
7. Editor's note: The facts of the matter were also documented in our time. See, for example, the study of the Egyptian peasantry by E. Taylor, in T. Shanin (ed.), *Peasants and Peasant Societies*, Oxford, 1982.

3

Credit in the Peasant Economy

In order to achieve a sufficiently clear and distinctive approach when investigating the forms of credit organization used by co-operatives in practice, we first need to become familiar with the role that credit plays, or can play, in the turnovers of the peasant economy.

Only after making a detailed study of every type of household, with respect to the make-up of its capital and of its turnover – that is, of the ways in which capital is laid out and replenished – are we able to judge how far the various conditions on which credit is made available are suited to this type of economic organization; and how far, in the conditions of this circulation, the sums advanced on credit can be repaid by the borrower. The material collected by expeditions to investigate budgets – undertaken both before the war and at the present time – as well as special calculations recently made, enables us to build up a fairly detailed picture as to the make-up of the capital of the peasant household and as to its turnover.

Table 5 gives us some idea of the extent to which manpower was equipped with the means of production. The figures quoted, like all figures derived from budget studies, provide us with very precise figures which visibly show the interrelationship between different components of basic capital. But it must all the same be borne in mind that budget figures are nearly always based on a record of the better educated households, which are able to provide detailed economic information about themselves. Because of this, they usually provide us with somewhat inflated figures in comparison with the mass average.

In order to correct this inaccuracy, V. A. Lipinskii and I made a calculation of necessary amendments and offered an assessment of

Table 5 *Value of the implements of production of the peasant economy calculated per household immediately before the war*

Districts	Value per household in roubles				Total value per hectare of land under cultivation
	Buildings	Cattle	Stock	Total	
Starobelsk, Kharkov province	420.5	471.2	100.4	993.6	103.4
Volokolamsk, Moscow province	909.0	268.0	189.0	1365.6	229.7
Vologda, Vologda province	453.0	137.0	82.3	672.3	257.8
Totem, Vologda province	313.9	108.9	44.1	466.9	126.3
Gzhatsk, Smolensk province	1123.6	212.7	83.2	1419.5	330.1
Sychev, Smolensk province	1262.0	174.0	100.2	1536.2	256.3
Porech, Smolensk province	1309.0	267.0	74.0	1650.6	275.0
Dorogobuzh, Smolensk province	717.0	271.5	82.0	1070.6	237.8
Voronezh province	341.0	130.2	79.1	652.2	68.2
Tambov province	550.5	316.5	98.1	965.1	148.3
Chernigov province	504.5	512.5	238.8	1255.8	153.2
Novgorod province	489.0	173.3	82.0	500.3	148.4

the means of production available to average peasant households according to the mass data of 1924. We have arrived at the following picture of the extent to which average peasant households in different provinces of the USSR were equipped with the means of production, expressed in money terms (Table 6).

Table 6 *Value of cattle, stock and buildings per peasant household (in gold roubles) (1924).*

Regions	Roubles per household	Regions	Roubles per household
Far North	668.4	North Caucasus	822.3
Northern	635.3	Crimea	907.3
North-Western	784.0	Kirghizia	674.3
Western	879.3	Siberia	675.0
Moscow Industrial	902.8	Far Eastern	674.4
Central Black Earth	578.1	Turkestan	696.8
Volga-Kama	849.6	Byelorussia	791.2
Urals	709.2	Transcaucasia	709.5
Bashkiria	560.4	Ukraine	691.4
Middle Volga	623.7		
Lower Volga	687.6	For the USSR	716.1

There is a certain variation in the figures which is due not only to the varying quantities of the means of production in households, but also to variations in the prices of the means of production in different areas, which are sometimes very substantial. Also, it is essential, when examining these figures, to remember that they are no more than arithmetical averages. Because of the demographic and social differentiation between peasant households, the extent to which they are provided with capital varies greatly. Thus, for example, according to an investigation of budgets in Novgorod (1910) the group which sowed on a small scale had 872.37 roubles of capital per average household, while the group which sowed on a large scale had 2,349.63 roubles. In the Starobelsk district in the Kharkov province (1911), the amount of basic capital per household varied between 455.75 and 2,925.02 roubles. Of particular interest are the results of the social analysis of the last Volokolamsk expedition of 1925 (Table 7).

Therefore, when we set out to analyse the organization of capital in peasant households, we have to bear in mind that in this respect, they represent large variations of magnitude. However, what is

56   *Credit in the Peasant Economy*

Table 7 *Value of household buildings, cattle and agricultural stock among different social groups of peasant households in Volokolamsk 1925*

	Value per household in roubles			
	Buildings	Cattle	Agricultural stock	Total
Proletarianized households	72.08	89.80	27.51	189.39
Semi-proletarianized households	219.53	165.51	54.49	439.43
Households based solely on their members' labour	361.74	287.20	146.91	795.85
Households mostly based on their members' labour	755.19	482.30	389.91	1627.40
Semi-capitalist households	1360.49	512.90	1145.23	3018.62

important to us in the present book is not so much the extent to which labour is equipped with capital as the conditions in which this capital circulates within the household.

The following excerpt from the calculation taken from the main volume of a book-keeping analysis of an average household in Volokolamsk enables us to gain an insight into the economic turnover of capital invested in buildings. When looking at the figures quoted, we must first of all note that the stock with which the household began the economic year was subject to considerable wear and tear during the year; and its value declined from 187.04 to 156.84 roubles (see Table 8). In order to replace the worn-out and disused implements in the stock, new stock was acquired which cost 12.04 roubles (the so-called capital repair stock). After that, the maintenance of the stock in working order required current repairs costing 10.82 roubles.

Thus in the Volokolamsk district the cost of annual purchases (renewal) of stock amounted to 6.4 per cent of its overall value within the household.[1] And current repairs came to 58 per cent of this overall value. In the Starobelsk district in the Kharkov province, purchases comprised 18.8 per cent and repairs comprised 5.2 per cent of the initial value of the stock.

In compiling these calculations, we should also note the expenditure on small items of stock which became worn out during the year

Credit in the Peasant Economy 57

Table 8 *Calculation of stock of an average household in Volokolamsk*

Account to which credited	Debit	Roubles	Roubles	Credit	Roubles	Roubles	Account to which debited
Capital	Value of stock at beginning of year	187.64		Value of stock at end of year	168.88		Capital
Cash	Purchase of new stock	12.04		Deducted: for tillage 92%	44.20		Tillage
	Repairs	10.82		for meadow cultivation 8%	3.85		Cattle-rearing
	Small purchases	6.43					
Total:			216.93			216.93	

and which – like current repairs – relate to current expenditure and not to the turnover of basic capital. What is particularly important to us in the above analysis is the evidence of constant wear and tear of the basic capital invested in the stock. The rate at which such wear and tear occurs depends on what the implements consist of. The more the household possesses complex machinery which serves for many years, the lower this rate will be.

A similar picture emerges from an analysis of the existence of basic capital in buildings. Thus a calculation relating to buildings with regard to this same household in Volokolamsk gives us the following picture (see Table 9). The make-up of the debit side of the account is similar to the account relating to stock. However, because of the greater durability of buildings the annual average percentage of expenditure on erecting and repairing them amounts in all to 5.2 per cent of their total value. The corresponding figures for the Starobelsk district are 5.1 per cent.

The annual wear and tear and the annual expenditure on the maintenance and repair of stock and buildings, which we analysed above, represent the cost of the use of this capital and constitute an overhead cost for the various branches of the economy which utilize them. This cost, if calculated in relation to one *desyatina* [2.7 acres] of land and one worker, produces the following rates for the Volokolamsk district (Table 10).

We can see that because of the buildings' long period of serviceable use, the annual cost of their maintenance is no greater

Credit in the Peasant Economy

Table 9 *Inventory of buildings of an average household in Volokolamsk*

Account to which credited	Debit	Roubles	Roubles	Credit	Roubles	Roubles	Account to which debited
Capital	50% of value of hut at beginning of year*	218.18		50% of value of hut at end of year	209.45		Capital
	Value of services at beginning of year	472.72		Value of services at end of year	452.81		
			690.90			662.26	
Cash	Erection of new buildings	2.80		Deductions: for tillage 65%	9.25		Tillage
	Repairs	14.60		for market-gardening 0.5%	0.30		Market-gardening
	Insurance	14.32					
			31.72	for meadow cultivation 5.5%	3.32		Meadow-cultivation
				for horse traction 16.3%	9.95		Horse traction
				for productive cattle-rearing 12%	7.54		Cattle-rearing
						60.36	
Total:			722.62			722.62	

* Footnote: The other half of the value of the hut is set against the personal account of the family.

for the household than the cost of maintaining stock; despite the fact that within each household, the capital invested in buildings is considerably greater than the capital invested in stock.

Entirely different principles apply to the turnover of the capital which the household has invested in cattle, as can for example be

Table 10 *Value of usage for one year*

	Per hectare of arable land	Per hectare of meadow	Per worker
Stock	4.92 roubles	1.06 roubles	13.99 roubles
Buildings	4.32 roubles	1.09 roubles	18.50 roubles

seen from Table 11, which relates to an average peasant household in Starobelsk.

When we come to the last category of capital – turnover capital – we have to emphasize its unique position in a self-employed peasant family farm [*trudovoye khozyaistvo*]. First of all, its relative scale in a family farm is considerably more modest than in a capitalist farm, since it lacks the main item of expenditure which in a capitalist enterprise is derived from turnover capital, namely, wages. The remaining components, that is, seeds, manure and fodder, to a large extent complete their turnover within the household in kind; and only part of the expenditure takes the form of money transactions.

When attempting to ascertain the scale of turnover of capital, we have to isolate from the overall expenses and receipts those items which should be so classified. Thus, for example, in the Starobelsk district of the Kharkov province, the items shown in Table 12 may be included among the expenses defined as turnover capital.

Thus turnover capital which is replenished every year in full, even though it exists mostly in kind, nevertheless represents a very substantial magnitude in the life of a peasant household. According to calculations which we made for 1924, throughout the USSR for every 100 roubles of basic capital in the peasant household, the latter had shown 15.9 roubles of turnover capital (i.e. seeds, fodder, money expenditure, etc., but not counting the value of the labour of its family). At the same time, this figure fluctuated sharply, depending on what farming system was being adopted. It was lowest in the Transcaucasian region (8.4 roubles per 100) and in Turkestan (9.1 roubles per 100); and it was highest in the cattle-rearing regions of the far north (23.3 roubles) and of the west (22.8 roubles).

The capacity for capital formation varied greatly among different strata in rural areas. Thus, for example, for the Starobelsk district (1910) we have the data of Table 13.

It is therefore quite obvious that a credit transaction, and the possibility of repaying the loan, can be examined only in connection with the processes which we have described of the circulation of capital within the household to which credit has been advanced. This

Table 11 *Composition and movement of capital invested in cattle*

Groups classified according to area of sowing per household	Value of cattle at beginning of year	Sold for sum of	Died or disposed of	Slaughtered	Bought for sum of	Increase in value during year*	Off-spring valued at sum of	Obtained in kind from elsewhere	Value of cattle at end of year
				ROUBLES					
0.00	19.9	10.4	0.7	4.3	12.2	1.5	1.5	1.7	19.9
0.001–3.00	50.2	40.1	0.7	2.7	42.2	3.2	3.2	1.6	54.2
3.01–7.50	228.8	71.1	22.6	22.4	68.1	28.4	28.4	17.4	226.9
7.51–15.00	539.0	138.1	31.8	49.0	123.7	31.2	31.2	33.9	519.0
15.01 . . .	961.0	181.6	59.5	79.2	145.9	69.3	69.3	69.3	937.0
Average	412.4	100.7	26.8	36.1	88.9	32.4	28.6	3.3	402.0

* Footnote: This includes the value of animals which have grown as well as all other changes in the value of animals, ascertained by comparing their value at the beginning and end of the year.

Table 12 *Economic expenditure out of turnover capital in an average household in the Starobelsk district*

	In kind	In money	Total
Payments and obligatory services	–	12.54	12.54
Journeys to the market and to towns	–	1.71	1.71
Rent	–	30.92	30.92
Hire of workers	–	17.05	17.05
Hire of agricultural implements	–	0.94	0.94
Other general household expenses	–	10.37	10.37
Seeds for planting in the fields	47.40	–	47.40
Seeds for planting in market gardens	2.00	1.00	3.00
Fodder for cattle	214.5	4.24	218.74
Payments to shepherds	0.84	1.92	2.76
Shoeing of horses	–	1.01	1.01
Payments for work done	–	2.94	2.94
Grain and other food for poultry	9.47	0.54	10.01
Total per average household	274.26	85.18	359.44
Total per worker	77.0	23.86	100.86

idea, which might have seemed quite obvious and which was frequently applied in practice when credit was advanced, for some reason failed for a very long time to be reflected in economic theory. Credit was reduced to a simple trade in capital or, more precisely, to a trade in the use of capital, which was remunerated by payment of interest for this use when the loan was reimbursed. The creditor always assumed that his loan was secured by the property of the borrower; and even in the case of a loan granted on the security of goods, the lender in no way required that the money should be invested in the handling of this commodity. His sole interest was in the selling price of the commodity pledged.

The first attempt to provide a somewhat different interpretation, when analysing credit turnovers, was made by the Russian

Table 13 *Rates of reproduction of capital among groups with differing areas under cultivation (Starobelsk district)*

Area under cultivation (Desyatiny)	Expenditure on reproduction of capital		Expenditure on capital formation per 100 roubles of expenditure on personal needs
	Total	including money expenditure	
0.05– 3.00	87.31	31.32	48.8
3.01– 7.00	208.38	58.71	68.9
7.01–15.00	472.40	150.32	96.2
15.01–	894.29	309.62	112.7
Average	398.76	131.90	92.5

economist, V. Kossinskii; and what occasioned it was the study of co-operative credit in Germany.

V. Kossinskii started out from the assumption that capital advanced on credit to a borrower goes through a productive turnover in the latter's household in accordance with Karl Marx's classic formula: $D - T - D + d$, where at the end of the resulting cycle the sum of money D is returned to the borrower together with an amount of surplus value realized to the amount of d. Part of this surplus value represents interest on the loan, while the other part of the surplus value d is retained as the borrower's income and represents his capitalist motive for using the capital which he has borrowed.

According to this theory, the success of a credit transaction depended on the success of the cycle of capital circulation; and the loan itself was economically secured not by the borrower's property, which only came into the picture in the case of bankruptcy, but by the very success of the circulation of the capital obtained by the borrower, which made it possible to realize not only the value D but also the surplus value d. Therefore, according to Kossinskii's definition, credit was an example of 'the circulation of capital in an enterprise owned by somebody else'.

Kossinskii's theory, which made a powerful impression on the economists of his time, did provide a correct basic idea for elaborating the problems of credit; and it was immediately recognized as a basis for the granting of co-operative credit for the purposes of production. Its primary weakness was its too formal interpretation of Marx's formula. A reader who lacked knowledge of practical life might suppose that capital advanced on loan had the nature of some kind of mystical substance which, as it circulated through the cycle of production, continued to preserve its distinct identity; and that it would – always, in every case, automatically and without any effort by the particular borrower – end up by producing $D + d$.

In reality, however, the process of capital circulation is much more complex. Capital loaned for production (D) has – at the moment when it exists in the form of money – got a certain money value and continues to do so when it is converted into the initial means of production and raw materials, since at this stage all the means of production which are purchased are equal in value to the value of the money spent on purchasing them: that is, their value is determined by the current market situation, which reflects the social relationships existing in the country's economic life at the moment of the purchase. However, from the moment when the production processes begin, when the initial means of production and raw material cease to appear in their original condition and become half-finished products, they also become things and not values. They constitute the T of Marx's formula calculated, however, according to units in kind. If for any fortuitous reason the production process has been interrupted before the cycle is complete, the realization of all these 'things' on the basis of current market prices will never give us the full amount of the capital expended. The 'things' of our production process cannot acquire the valuation which concerns us as organizers of production until the production cycle has been completed – when we have finished products ready to be sold, that is, commodities which we have produced.

Their sale on the market (or market valuation) provides us with a fresh sum of money. But this sum cannot in any sense represent the mystically revived abstract substance of the original capital D, because the market valuation of the commodities obtained is not made in relation to the same commodities which were the yardstick for the original valuation; and, above all, it was made at a different moment in time and in an already different market situation, since the production process cannot assume a static national economy.

What happens from the market's point of view is not the restoration of the value of the old capital $D1$ in terms of book-keeping records and calculations, but the creation of new money capital $D2$. If the product which we have produced remains as socially necessary in the country as it was at the time when we drew up our production estimates, and if the production costs incurred were no higher than those currently dictated by the interplay of socially necessary economic factors, then the system of market prices guarantees that, in our particular case, D_2 will be greater than D_1; that the sum of D_2, when realized, can be divided into D_1 and d; and that D_1 can again be invested in a new production cycle.

Thus, when examining the process which is, not quite correctly, termed the process of capital circulation, we have the following series of conditions on whose successful fulfilment and appropriate combination the success of the entire production cycle depends.

1. The availablity of a certain money capital, D_1.
2. The acquisition – out of the sum of money D advanced for the production of capital – of the necessary means of production and raw materials; and the hire of manpower at prices which do not exceed the estimates of profitability.
3. The use, in the process of production, of the 'things' which have been bought and of labour with the technical skill and success needed to ensure that the cost of the end product T is no greater than its socially necessary cost.
4. The sale of the end product at currently prevailing prices corresponding to the social conditions in the country's economy at the time when the manufactured commodities are sold; and the formation of a new sum of money D_2.
5. The allocation – out of the sum of money D_2 obtained from the sale of the commodity – of fresh capital to be advanced for the continuation of production at the same level. This will be on the same scale – D_1 – if market conditions are unchanged; or D_3, in the event of conditions having changed.
6. The division of profits (on the principle $d = D_2 - D_1$ or $d = D_2 - D_3$) between the sum which the owner takes out of the enterprise and the sum which he allocates for the expansion of the farm's capital.

Marx's old formula has been translated, in the above case, from static into dynamic terms. When modified in this way, it really can, as V. Kossinskii rightly supposes, prove invaluable for an understanding of the credit turnover and its importance in the outlay and renewal of capital circulating in production. It is, however, necessary to note certain distinctive characteristics which the process of circulation we described acquires in the case of market-oriented peasant households which do not use hired labour.

All work in such a household is performed by the manpower of the family itself. This means that wages – actually paid out at levels spontaneously and objectively determined by the labour market in accordance with the social relations currently prevailing in the country's economic life – do not exist as a real economic phenomenon in a peasant family household. Expenditure on personal needs by those working in a peasant household represent not the spending of an objectively determined wage, but the earmarking for the given item of a part of the household's overall income, which is also used to finance the renewal of capital. Because of this, the process of capital renewal in peasant family households is linked to the level of well-being attained in the given year.

Indeed, if we are speaking of real economic phenomena, the gross income of a capitalist farm can be broken down into the sums spent on the renewal of capital (renewal of the means of production and of the wages fund) and pure profit. If the production cycle has turned out to be so unsuccessful that gross income not only fails to yield any profit but fails even to achieve the renewal of capital, then the entrepreneur, unless he can draw on credit, is obliged to curtail the renewal of his capital and thus curtail production itself.

The situation is somewhat different for a market-oriented peasant family household, whose gross income divides into three parts: 1. The value of the renewal of the material means of production, as objectively determined by the market; 2. the family's personal consumption; and 3. sums set aside for the purpose of capital accumulation.

In the event of a fall in gross income – as in the case of a farm organized on capitalist principles – one of the first results will be a reduction of capital accumulation. In the event of a further drop in gross income, a family farm, unlike a farm which pays wages, can continue to maintain the volume of activity at the old level, by cutting personal consumption so as still to pay for the complete renewal of the material part of its capital. As is shown by a statistical analysis of budgets,[2] this renewal of capital at the cost of reducing consumption continues for a fairly long time; but it does so only up to a certain point, beyond which there is a parallel reduction in both cases.

This special characteristic of the peasant household considerably enhances the ability of peasant households to renew their capital and makes it easier for them to recover from economic crises.[3] We must, of course, take this, too, into account when examining the forms and conditions for the granting of credit to peasant households.

Another, and no less important, special characteristic relating to the renewal of capital in peasant households arises from a further fact which stems from the small scale of the enterprise.

In the preceding tables, we showed the rates of wear and tear and of the corresponding renewal of basic capital. In terms of national economic aggregates, involving hundreds and thousands of households, these rates have the actual significance of actual expenditure continuously undertaken every year. A thousand peasant households possessing, let us suppose, 2,000 ploughs may be expected, according to the law of large numbers, to discard and replace 250 ploughs every year; this proportion remains significantly stable from year to year and corresponds to the percentage of amortization.

In large farms, which comprise hundreds of items of stock, annual expenditure on the renewal of capital will, by virtue of the same law of large numbers, also be relatively constant and continuous from year to year. But we find something different in the case of the small family enterprise whose items of stock consist of only a few units; and where the annual rates of renewal (amortization) are expressed as proportions of an item. In this case, the discarding of old items and the acquisition of new ones is far from being an annual occurrence. The renewal of capital over a number of years may be equal to zero; and subsequently, in some one particular year, it is reflected in figures which are many times higher than the normal percentage for amortization, even in large farms.

This creates difficulties for the peasant household, particularly when one takes into account the inherent link already pointed out between the process of capital renewal in peasant households and their personal budgets. During years when none of the items of the means of production requires replacement, the peasant household usually sets absolutely nothing aside 'for capital amortization' and consumes the part of the income defined for this purpose by the relations between the market prices. In years when such replacement becomes unavoidable, the scale of the expenditure on renewal is many times greater than the percentage of amortization and represents a substantial and burdensome deduction from the earned income and, therefore, from the consumer budget.

This is why the process of capital renewal is much more difficult for a peasant household than for a large-scale privately-owned enterprise, and why it is, to a considerable degree, more dependent on the availability of credit or the postponement of payments in order to convert renewal at irregular intervals into more evenly spaced annual payments.

Summing up our analysis, we have to regard the process of circulation as normal when it entails a fully completed process of capital renewal at the centre of all economic activity. V. Kossinskii's formula that 'credit is the circulation of capital in an enterprise owned by someone else' has to be supplemented, by pointing out that the success of this circulation depends on the process of capital renewal being fully completed by the end of the economic cycle, and within the period for which the loan was granted; and since the credit is in the form of money, the renewal must have been carried to the stage where this capital can be converted into money.

Up to now we have been discussing ways of maintaining a household's capital in an unchanged condition, by constant renewal. However, a peasant household never remains in an unchanged condition, but is constantly altering its scope, either because of changes in the state of the market or because of changes in the composition of its family. Therefore, in addition to the task of renewing capital, we come up against the problems of expanding it. It should be noted here that the need for such expansion often arises not only because of the organic growth of the family and the household, but because of all kinds of changes in the organization of production techniques.

The transition from manual threshing to the threshing machine, from the use of cattle for manure to their productive use, etc., all these changes make it necessary to improve the way labour is equipped with the means of production; and make it necessary in the case of a market-oriented capitalist farm to improve the organic structure of the capital.

In other words, it becomes necessary to increase the farm's capital not only in relation to the farm taken as a whole, but in relation to every individual unit of the farm's manpower. Such an improvement in the way labour is equipped will itself serve a purpose only when, as a result of this, the family's manpower begins to earn a more abundant livelihood and/or gets a greater opportunity for capital accumulation. This increase in income, as a result of the household becoming more capital-intensive, subsequently makes it possible not only to finance the renewal of the part of the capital which has been freshly created, but also to recoup the original outlay.

These are the considerations that determine the profitability of investing fresh capital sums in the economic turnover. It is perfectly obvious that after an estimate has been made of the profitability of making new capital outlays, the basic question then arises: *where are the funds for this increase to be obtained?*

We must acknowledge that, in this case, the capital increment can be obtained by the same means as outlay on ordinary capital renewal – that is, by a deduction from income. Here, the deduction for growth has to be added to the normal deductions for the renewal of obsolete items of capital. In most cases, the peasant household, particularly in years when the harvest is good, is in a position to make such deductions, sometimes even at the cost of slightly curtailing its consumption. But cases are quite common when the unsatisfied needs of consumption are so acute that they swallow up the entire income earned by the family's labour, leaving no opportunity for deductions from income. In that case the household is left with only one way of enhancing its capital – by using credit.

From the foregoing analysis, we can see two cases where credit is relevant to a peasant household:

1. The case where owing to the unevenness of the process of capital renewal in the peasant household, the current replacement of means of production involves a painful cut in the family's food budget; and where, thanks to the availability of credit, a large item of one-off expenditure is spaced out over a number of years and is converted into a series of redemption payments with interest. The loan is repaid in later years out of the ordinary general fund for capital renewal. However, the period for repayment does not necessarily have to be the same as the period for amortization, but can be shorter. In this case the role of credit is in effect to maintain the household's capital.
2. The case where, for one reason or another, the household needs to increase its capital, but is unable to pay the initial outlay through deductions from the household's usual income earned by its labour. In this case credit enlarges the household's capital; and both the loan and interest are repaid out of the increase in the household's capacity for capital formation.

There are also certain other possible cases which are encountered in practice in the peasant economy. There is, for example:

3. The case where the borrowed capital is invested in a productive commodity which does not depreciate during the production process, but remains, so to speak, in use and does not lose its value; as a result of which the loan is repaid by selling this commodity on the market. (For example, where a peasant borrows money to buy a cow and then, having profited from its milk during the season, sells the cow and repays the loan out of the proceeds of the sale.) In this case, it is only the interest on the loan which is paid out of the funds for capital renewal.

Finally, a very special situation arises in:

4. The case where the funds borrowed are not used for production but are spent on the needs of consumption; and where the debt is subsequently repaid out of the household's general income, thereby reducing the household's capacity for capital renewal.

When studying the money economy of the peasant homestead, we can nearly always note periods of an acute shortage of money and periods when the household is comparatively well provided with cash. This phenomenon, which can be observed in budgets where there is no shortage of money at all, is the result of the fact that the money receipts and expenses of a peasant homestead are by no means evenly spaced out over the months of the year, nor do they dovetail with one another.

In order to grasp this aspect of the role played by money in the peasant homestead, it is enough to look at Figure 2 – based on the results of a special questionnaire – which shows how money receipts and expenditure in peasant households in the Moscow province were spaced out over time.[4]

A good picture is also provided by data on the Chernigov province in a book by N. I. Kostrov, which shows the following distribution of money payments and receipts for each third of the year, expressed in percentage terms:

Table 14

	First Third	Second Third	Third Third	Total
Expenses	27.1%	46.9%	26%	100%
Sales	5.8%	15.2%	79%	100%

Figure 2: Movement of money resources during different months – Moscow province

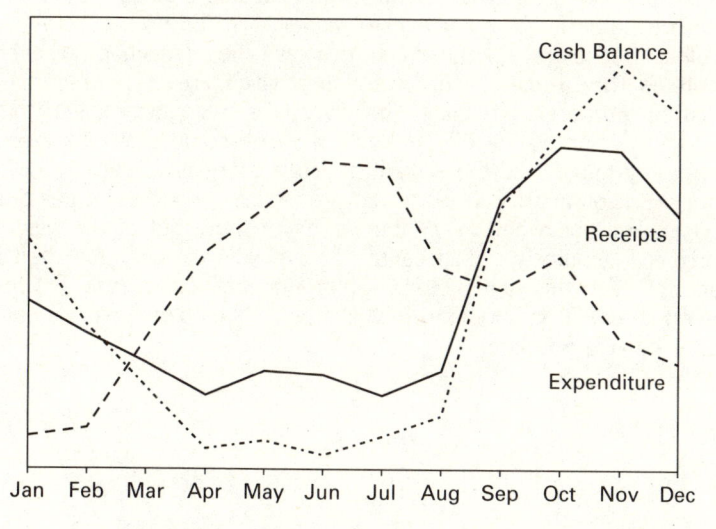

We can see that expenditure is at its highest in the spring months and that receipts are at their lowest in the autumn months. This discrepancy gives rise to a whole number of special credit operations which we could call balancing operations, since they enable the peasant, without disturbing his economic system, to balance money receipts and expenses which arise at different times.

Let us suppose that in May the peasant needs to incur expenditure of 25 roubles. He might cover this by selling a couple of calves and fetching for them a price of some 30 roubles. However, this sale cannot be considered advantageous, since in the following September these same calves could be sold for 70 roubles. It is clearly more advantageous for our peasant to finance the 25 rouble expenditure through a loan which can easily be repaid together with all interest by the autumn sale, which would also yield a considerable profit over and above the repayment of the debt.

We can, incidentally, bring balancing credit within the classification of the types of credit examined above. But in order to do this, we have, notionally, to divide the personality of the householder into two; and to see him as buying goods from himself out of borrowed money in the hope of making a speculative gain from this purchase as a result of the expected change in price. This kind of interpretation is formal and tentative, however, and it can add nothing to our understanding of the process.

Such are the uses of credit which can be observed in the peasant household. It is not hard to see, after analysing them, that the peasant household is fully capable of making a proper and timely repayment of credit advanced to it, on condition that the resources which it obtains are productively channelled in the proper way into profitable economic operations. In this case the production turnover provides the creditor with guarantees which are no worse than secured property. It must, however, be recognized that this 'productive purpose' is of value only where the conditions on which credit is granted, and the ways in which it is organized, are adapted with precision and flexibility to the production processes of peasant households; and when every item of expenditure out of the capital borrowed genuinely enhances the household's capacity for capital renewal. All this undoubtedly confronts the organizers of co-operative credit with an extremely subtle and complex organizational and economic problem.

NOTES

1. For reasons which we do not know, expenditure on purchases was three times less than the norm, with the result that in this year worn-out capital was not fully replaced.
2. For evidence see A. Chayanov, *The Theory of the Peasant Economy*, Madison, 1982.
3. It was precisely for this reason that the celebrated agricultural crisis in Europe at the end of the nineteenth century was more easily endured by peasant households than by capitalist agriculture.
4. A. Chayanov, *Opyt anketnogo issledovaniya denezhnykh elementov krest'yanskogo khozyaistva* [The Experiment of a Questionnaire Survey of the Pecuniary Elements in the Peasant Economy], Moscow, 1912.

4
Co-operative Credit Societies

Co-operative credit societies evolved over many decades. They evolved independently of the influence of those subtly elaborated theoretical premises relating to small-scale credit which we now have and which were developed by scholars in recent years, at a time when all the most important foundations of co-operative credit societies had already become finally established. Spontaneously and gradually, without any consciously conceived plan of construction, they developed their own principles and traditions, in practice moving forward from the concrete solution of one particular problem to the solution of another.

Through the gradual selection of viable principles of organization and forms of work, and through the demise of other forms which were inapposite, there emerged those basic Principles of Raifeizen, which constitute the foundation of the entire structure of co-operative credit societies.[1] It should at the same time be noted that most of these basic principles were conceived by their original authors for an entirely different purpose and for the solution of problems which differed from those that had to be solved in reality.

We saw in the last chapter how acute is the peasant household's need for credit; and we also know on what foundations such credit can be based. It is now necessary to examine what kind of credit apparatus can help to make this necessary credit accessible to the small family producer.

With regard to the existing forms of granting credit we can, with reasonable clarity, divide credit transactions into two large groups. On the one hand, we can observe the extensive area of credit relationships which are organized as their own kind of international capital market. Bank credit, industrial and commodity credit, credit based on bills of exchange, with its special system of stock-taking – all this constitutes a dense network of credit relationships which absorbs unused capital and puts it into the hands of economic organizers who need funds. On this market, capital is a depersonalized commodity which has a uniform price for all similar credit transactions, in the form of a rate of interest on loans. As a rule, owing to the fact that a considerable volume of capital has been accumulated, its price – that is, the above-mentioned rate of interest on lending – usually proves to be low: it has been varying from 3 to 7 per cent per annum.

In addition to this type of market credit, there exist very numerous forms of credit which is unorganized in the sense just mentioned. The small peasant producer, who works far from the centres of market credit and who is unable to satisfy many of the conditions which credit centres usually impose upon their clients, is almost always denied the opportunity to avail himself of cheap organized credit. He meets his inherent need for funds from local sources, from the unused capital at the disposal of his more prosperous neighbours. Credit of this kind remains unconnected with the organized capital market; and the conditions of this market, including the low rate of interest on capital loans, have no influence on it.

The acute need for funds and the small volume of locally accumulated capital drive up the price of capital to an extraordinary degree and give rise to so-called 'kulakism', that is, to a kind of trade by the local rich who provide money on credit at interest rates of 30–50 per cent or more per annum.

Therefore, for a number of reasons which we shall examine below, there still exists, alongside the international capital market, a system of local credit in the countryside based on usury, which, being unconnected with the first type of credit, is not subject to its laws. The result has been that whereas capital was loaned by the local bank at interest rates of 6–8 per cent, a peasant who lived several miles away from the town was obliged to pay monstrous interest rates.

This disparity has long since attracted the attention of public figures and statesmen; and the question was more than once raised as to whether it would be possible to destroy usury and provide the peasant with access to the global capital market with its cheap credit. However, at the beginning of the nineteenth century numerous efforts to solve this problem had not led to positive results; and only over the course of time did it prove possible to devise feasible methods of linking the peasantry with the credit system as a whole.

In order to understand the significance of these methods, which were realized by means of co-operative credit societies, we need to clarify, first, the obstacles which had previously prevented the peasant household from making use of the general credit apparatus. Here, the main hindrances arose from two circumstances, namely:

1. The remoteness of the peasant household from the apparatus of banking; and
2. The small size of the peasant household itself and, therefore, the smallness of the amount of credit which it needed.

These circumstances made it extremely difficult, in the first place, to verify the solvency of a peasant seeking a loan. The compilation of reference documents and the sending of an agent to the locality might cost the bank more than the amount of the loan being requested. The overhead expenses involved in obtaining securities for loans might be so high that they might easily, in relation to the loan, exceed the highest rates of interest charged by usurers.

Even greater difficulties faced the banking apparatus in the event of non-repayment of the loan; since the costs of recovery could not, of course, be met in any way out of interest on the loan.

If one adds to this the fact that in the eyes of the capitalist credit system the only possible security for a loan was the capital or property of the borrower – which, in the case of a peasant family, was negligible and difficult to sell off – then we can understand why, for city banks, the granting of credit to peasant households was not only not very advantageous but was simply impossible.

It was obviously necessary to find a credit apparatus which would be in the immediate vicinity of the peasant household, which could observe it at every moment and do so cheaply, and which would base its credit operations not on secured property but on some other kind of security. Such a credit apparatus was realized in the credit association created according to the principles laid down by Raifeizen.

What it fundamentally represented was an alliance of peasant households for the purpose of jointly obtaining credit. The consolidation of the demand for credit of a large number of households multiplied the volume of credit to such an extent that all overhead expenses required for verifying the solvency of a collective borrower and for the recovery of loans in the event of their non-payment were brought down to the usual, relatively low levels.

Taking into account the ideas concerning credit within the capitalist credit system, Raifeizen based his credit alliance of households on the principle of the mutual and reciprocal undertaking of all members to pledge their entire assets to underwrite the alliance's commitments. An undertaking of this kind was very impressive even in the eyes of a commercial operator – since the 300 or 500 peasant households comprising the alliance could provide a security of several hundred thousand roubles which could be realized in the event of bankruptcy.

We are inclined to think that the institution of unlimited liability was merely a concession to the ideology of the capitalist credit system since, in our opinion, the solvency of a credit association is established not by this, but by other principles of Raifeizen. And if we still support this approach at the present time, we do so only in order to get the members – through the threat of unlimited liability – to adopt the most active attitude possible towards the affairs of their co-operative, to monitor its work and to create a public opinion in the countryside which favours the meticulous repayment of loans.

Among the principles conducive to repayment, prime importance attaches to the proposition that all loans advanced by a credit co-operative should be advanced only on condition that they are used for the productive needs of the farm.

The productive purpose of the loan, as we saw in the last chapter, will itself provide the sources for the loan's repayment. Money which is channelled, on the basis of a well-founded economic calculation, towards the expansion of the peasant household's capital or even towards its mere renewal or towards the regulation of the household's money economy, will strengthen the peasant family's productive capacity and enable its workers more fully to develop their working energy and to increase the income which the family earns from its labour. And it is the increase in income, as we saw in the last chapter, which makes it possible to repay the loan advanced.

We must, admittedly, recognize that the very term 'productive purpose' lacks precision and allows of very wide interpretations, particularly when the loan advanced is used in order to regulate the money turnover. Can one, for example, describe expenditure on a wedding, or the purchase of flour when it is in short supply before the New Year, as productive expenditure?

In the practice of co-operatives, such cases have long since been included in the list of productive expenses. But from the point of view of theory this can be done only with some difficulty, by arguing that a wedding provides a peasant family with fresh manpower thus increasing its potential; and that expenditure on flour would, in a capitalist farm, count as expenditure on food, that is, as a part of wages and therefore as productive expenditure. We could say that in the broad sense of the term, productive expenditure should be taken to mean all expenditure which has an economic purpose and which may turn out to have involved no losses.

We accept the view that a credit association cannot be guided purely by formal considerations or by the formal attributes of productivity but that it has to grasp the economic essentials in each case. Sometimes the erection of a barn may turn out to be an unproductive expense, while the acquisition of a sheepskin coat (for example, by a cart-driver) may be entirely productive. The whole question hinges on the economic calculations upon which the proposed expenditure is based and on how far they offer a guarantee that the household will be able to recoup the capital advanced.

The productive purpose of loans was treated by Raifeizen as the most important framework for the theory of co-operative credit. Indeed, on more profound reflection, we are bound to admit that all the remaining 'principles' stem from this basic principle and represent the conditions on which the basic principle can be fulfilled. Thus it is self-evident that the productive purpose of loans can be guaranteed only when a loan is made available and spent under the constant observation of the institution which makes the credit available. This institution must be informed as to the state of the borrower's household and as to the use to which the loan advanced is being put. It is obvious that these two conditions will be fulfilled only if:

1. The borrower is known to the management and to the other members of the co-operative; and
2. The borrower's household is observed by the co-operative and is accessible to observation.

The first requirement gives rise to Raifeizen's third principle, according to which a credit association can grant credit only to its own members; and this in turn leads on to the fourth principle which requires that the geographical area of the association's activity should so far as possible be *restricted* or, to use the customary co-operative term, localized.

This last principle confronts us with the complex problem as to the limits of this localization.

It is obvious that the greater the area of activity, the greater will be the volume of credit turnover and the lower – in relation to every rouble of credit advanced – will be the overhead expenses on the administration and management of the association. But it is no less obvious that the more the area of its activity expands, the more remote the association will become from its borrowers; and it will, owing to the weakening of these ties, face an increasing number of loans which cannot be repaid and mounting expenses for the recovery of loans. Therefore, if the area of activity is too narrow this

Figure 3: Influence of the radius over which service is provided on costs

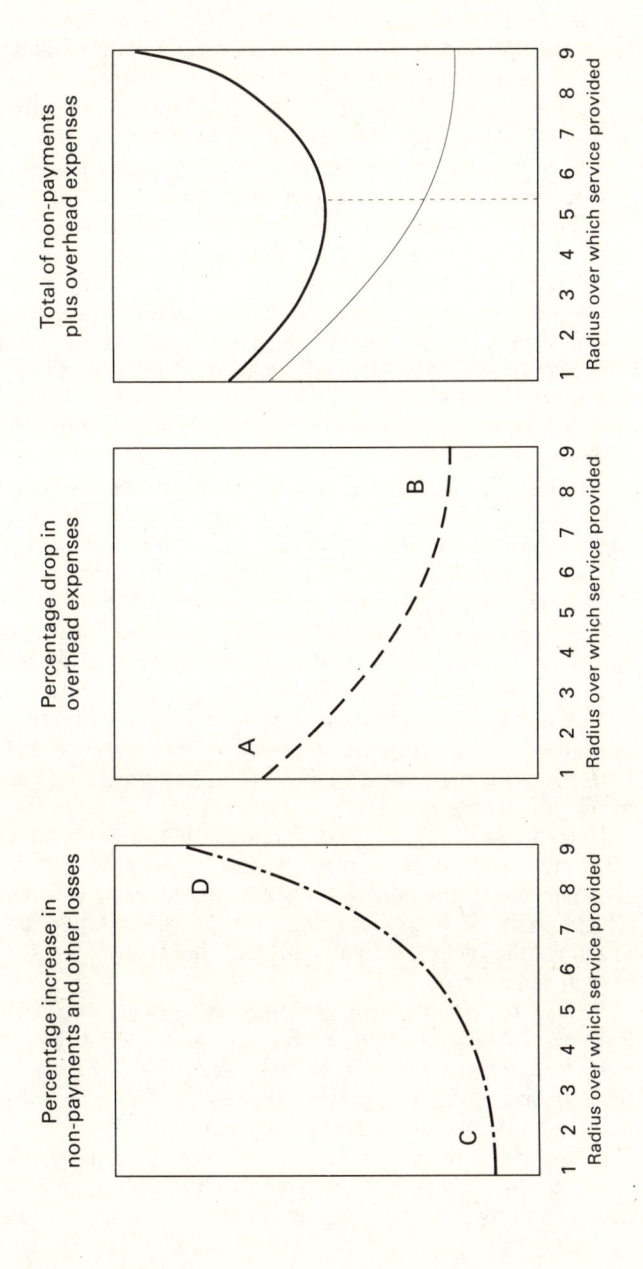

Percentage increase in non-payments and other losses

Percentage drop in overhead expenses

Total of non-payments plus overhead expenses

is a bad thing owing to the extremely high overhead costs, while if the area of activity is too wide this is also a bad thing because of the losses resulting from non-payments. What is essential is to search for some optimum area of activity where the sum of overhead costs and the sum of normal non-payments are minimal.

Figure 3 clearly shows how this optimum scale is arrived at. The curve AB represents the decline of overhead costs per rouble of credit, depending on the radius of the association's area of activity, that is, on the increase in the volume of credit turnover. The curve DS shows the mounting losses – as this radius increases – which result from postponements of payment or non-payment – calculated per rouble loaned by the association. The curve MR shows the sum of overhead costs and losses. As can be seen from the sketch, this is at its lowest at the point which corresponds to a radius of 5 kilometres, i.e. the optimum radius for the association's area of activity.

The shape of these curves and, therefore, the extent of the optimum radius, depends on a whole number of local economic conditions. The higher the population density in the area where the association operates, and the greater the financial and credit turnover in the local households, the more rapidly the overhead costs will fall as the association extends its area of activity and, therefore, the shorter the optimum radius will be.

It is interesting to note that the Russian credit associations, whose radius of activity was – under the policy of the old authorities – always considerably above the optimum, tried to obviate losses due to non-payment by imposing severe restrictions on the credit they advanced. They made it depend on the distance between the households of their members and the place where the association's management was situated.

Unfortunately, we have very limited factual material for ascertaining this optimum by empirical means. Research carried out by N. Makarov in the Moscow district provided us with indirect data which point to a gradual decrease in co-operative activity, the greater the distance between the households and the credit association's base.

We get a particularly vivid picture in regard to the association in Rzhev-Savelovsk. It can be seen from Table 15 that the further away the members live from where the association is based, the smaller the proportion of peasants who take part in it and the fewer the links between the association and its members.

When trying to provide rural areas not only with credit but with cheap credit, a co-operative credit society must do all it can to make the work of its own apparatus as cheap as possible. One way of doing

Table 15 *The effect of distance on local links with the credit association in Rzhev-Savelov, 1908–1910*

Distances	% of farmsteads participating in the association as members	% of members who did not take loans	Average size of loan per borrower	Number of payments on loans per member per year
0–4 versts	77%	44%	59 roubles	2.3
4–6 versts	59%	63%	23 roubles	1.8
6–16 versts	29%	83%	26 roubles	1.5
16 versts and over	15%	84%	15 roubles	1.2

this in the Raifeizen system was by offering the services of the association's management free of charge to its members. In a small-scale association which had a small credit turnover and which usually operated for two or three hours once or twice a week, the obligations of a member of the management were not burdensome. Raifeizen regarded these obligations as being of an honorary nature which meant that such work was unpaid, thus making credit considerably cheaper. Raifeizen accordingly based his idea of a credit association on the following five principles:

1. The unlimited mutual liability of members for the association's obligations.
2. The productive purpose of loans advanced by the association.
3. The granting of loans only to the association's members.
4. A small area of activity for the association.
5. The honorary and unpaid character of administrative work within the association.

Let us now try to clarify the way in which the apparatus of credit co-operatives has been built up on these foundations. We shall not deal with the managerial organs of co-operatives – the general meeting, the board of management and the observers' council – since their organization and work are widely known, and it is enough to consult their statutes or any popular pamphlet in order to become thoroughly familiar with them. What is much more complex is the financial structure of a co-operative and its financial methods.

The procedure for granting credit in agricultural associations which engage in credit operations is usually the following. Every member of the association wishing to avail himself of credit provides the board of management of the co-operative with information concerning himself and his household, the number of buildings which it contains, its stock, cattle and area of tillage. This report has to be verified and on the basis of this – and also of an assessment of the peasant's personal qualities, his capacity for work, enterprise and conscientiousness – he is granted a credit facility, that is, it is decided up to what limit money can be loaned to this comrade without risk to the association. Before the war the average amount of a credit facility for an association member was about 80 roubles.

If a credit facility is granted to a member of a co-operative, then he can, in case of need, request a loan from the board of management, indicating in his application the purpose of the loan, its amount and the period of repayment. Its purpose must be related to production and must not involve loss, its amount must be commensurate with its purpose, but must not, so far as possible, exceed the amount of the credit facility. The period of repayment is not more than six months. (Loans for longer periods are advanced only under a special procedure and provided that the association has special capital for long-term credit.) If the association has ready cash available and if the request is well founded from an economic point of view, the loan is advanced either for the whole or part of the sum requested; but with the prior deduction of interest covering the period for which the loan is requested.

If, let us suppose, a peasant receives a loan of 100 roubles for six months at an interest rate of 12 per cent per annum, then he will be given only 94 roubles while becoming indebted to the amount of 100 roubles.

Loans are made subject to three kinds of security, namely:

1. Personal confidence in the member who borrows;
2. A guarantee in the borrower's favour by some other member; and
3. The security of goods or cattle.

The sums advanced on the basis of personal confidence are comparatively small. If the loan requested exceeds this sum, it is advanced only where repayment is guaranteed by some other member of the association. Where this guarantee has been given, then in the event of non-payment steps are first of all taken to recover the money from the debtor; but if he is unable to pay, then the money is recovered from the guarantor. It must be noted that when a credit facility is granted to a member of the association, a decision is made not only as to the maximum amount which can be advanced to him on the basis of personal confidence and on the basis of a guarantee, but also as to the maximum amount which the guarantor can guarantee on behalf of others.

When an application is made for a secured loan, the amount advanced is related to the value of the security and is usually worth no more than three-quarters or two-thirds of its value. Acceptable securities may include agricultural products – grain, flax, animal hides, etc. – or cattle owned by the borrower.

In the first case, the products offered as security are usually handed over into the keeping of the association. But in cases where cattle are provided as security, they are left in their owner's possession but are made subject to a 'prohibition'; that is, the owner is deprived of the legal right to sell them or give them away or remove them without the special permission of the co-operative that has granted the credit.

The loan itself may be advanced either in money or in kind, i.e. in the form of an authorization to obtain goods from one of the association's agricultural warehouses. The handing over of agricultural machinery, manure, seeds, etc., is not, in this case, a violation of the Rochdale principle which allows trading only in cash, since the corresponding sum of money is immediately transferred by the association's credit department to the account of the warehouse, which represents a cash settlement. In this case we simply have the fusion of two kinds of operation into one: those of purchasing and of granting credit.

Once the loan has been granted and the borrower has given the association a promissory note, the association's board of management has the right to verify whether the money advanced really is being used for the purpose for which it was requested; and in the event of dishonesty by the member, it can demand the immediate reimbursement of the loan and expel the dishonest borrower from membership of the association.

If the economic turnover, for the purpose of which the loan was obtained, has not been completed by the time stipulated for repayment, or if the calculations of the householder who took out the loan have not been fully justified, he may ask the association's board of management for payment to be deferred; and this may be allowed – after careful examination of the circumstances and of the grounds for the request – usually for a period of no more than six months.

In cases where the borrower delays repayment for several days without having given notice to the board of management, a special fine is imposed on him for every day's delay.

Such are the methods of credit operations in the co-operative credit system. But where does the association obtain the resources for granting loans to its members?

The association's resources are made up of:

1. The association's basic capital;
2. Its reserve capital;
3. Its special capital and, in particular, its capital for long-term credit;
4. Money borrowed by the association for varying periods;
5. Money deposited with the association, subject to various conditions, by members of the public;
6. Money kept by the association for varying periods;

Let us examine each of these sources separately.

Basic capital may sometimes consist of shares held by members of the association. But according to Raifeizen's principles such capital is usually borrowed in the form of a long-term loan which is gradually paid off out of annual deductions from the association's profits. Basic capital which is raised in this way will, after the association has been operating for some years, increase, as public capital accumulated in the process of the association's actual work.

In the USSR at the present time, no final system has yet been devised for raising basic capital for the purpose of organizing small-scale credit. In all probability, however, the Central Agricultural Bank of the USSR will be able to undertake the commitment to finance the basic capital of agricultural associations – relying, in its local work, on local agricultural credit associations and local co-operative alliances.

An agricultural bank, as the centre of all agricultural credit, must in general devote a good deal of its effort to making credit of all kinds available to peasant households through their co-operatives. Its first priority must be to make credit available to local associations so as to finance their basic capital – on the one hand, because of the simplicity of this work and also, on the other hand, because only this bank, which can rely on large-scale and long-term state resources, is capable of performing this task on a mass basis. In all probability, the financing of agricultural credit will attract some of the resources of state savings banks as well as some insurance capital.

The association's *reserve capital* is formed gradually as its work proceeds by money set aside out of profits; it serves to underwrite all the association's commitments and is a means of covering possible fortuitous losses.

Special capital, created for various particular purposes, is raised through special borrowings, levies or donations and, since it forms part of the association's assets, it can also be used for granting credit. The purpose of providing credit may be directly served by special capital for long-term credit, the raising of which is exceptionally important in view of the acute need for long-term credit in our rural areas.

Borrowings are made by the association in the event of a shortage of resources from elsewhere: they are usually made for short periods from other co-operative organizations, from local banks or even from private individuals. When matters are properly handled they should not comprise a large part of the association's resources, being the most disadvantageous and the most expensive way of obtaining money.

The main channel of resources for co-operative credit must come from the population itself – through the transfer of their available resources to the association in the form of *deposits*. These can be made either by the association's members or, equally, by any local inhabitant who wishes to do so.

Provided that the co-operative movement has fully developed and is generally trusted, and provided that the population enjoys a perceptible measure of well-being, the inflow of deposits into co-operative associations is usually so great that it fully matches the association's credit operations.

A population which has become convinced of the stability of co-operative organizations will usually lend – at a low rate of interest – the money it has saved 'for a rainy day' and which it had previously kept in money-boxes, stockings and trunks. After that it will deposit the unused cash which it was for some reason unable to invest in an advantageous way and, finally, it will temporarily deposit its turnover capital when the latter, owing to the dead season, remains uninvested in production over a period of several months.

In concluding our description of co-operative credit, we need to dwell on certain extremely important aspects of such credit operations.

1. When describing the association's lending activities, we have said nothing about how the rates of interest on loans granted by the association are fixed. Nor could we do so without first examining where the association got its resources from. But we can now point out that the level of interest rates on loans is determined by the level of the rates of interest at which the association can obtain resources, either by borrowing or through deposits. Having obtained resources through deposits at a rate of interest of, let us suppose, 8 per cent per annum, the association will add on a further 2 or 3 per cent to help cover the costs of its apparatus and to make a profit; and will therefore offer loans at interest rates of 10 or 11 per cent per annum.

The difference between the rates for loans and for deposits in a well-run association must be as low as possible. The profits and the resources for the maintenance of the apparatus must be increased by increasing the credit turnover.

2. When granting loans it is always necessary to ascertain not only how much ready money is available but also for what periods it has been advanced to the association. Money which has been deposited for four months can under no circumstances be loaned for seven months, because when the time comes to repay the deposit there will be no possibility of getting this money from the borrower. In short, the periods for which loans are granted must always be more or less dovetailed with the periods for the repayment of deposits and borrowings, since any discrepancy in this respect can put the association in an exceptionally difficult position.

One can see that the importance of co-operative credit is immeasurably greater than mere help given to individual households in their work. As the co-operative credit system develops and grows stronger, it inevitably attracts, in the form of deposits, all the unused money in rural areas. By supplementing them with the resources and capital of the state, obtained through banks, it makes credit cheap and accessible to every peasant, and makes it generally available. As is now the case in industry, most of the circulating capital in agriculture will be borrowed, and borrowed through co-operatives. It is this borrowed co-operative capital which would be used to build up and organize butter manufacturing and potato-grinding factories, centres for stock-breeding, for the handling of machinery and for grain-cleaning, mills and other co-operative buildings; and this same borrowed co-operative capital would form the basis for all marketing, purchasing and reprocessing operations. In other words, if all the operations just enumerated acquire a wide scope, then they will lead to the gradual co-operative socialization of all capital circulating in agriculture and on the market for agricultural products.

It is customary for us to describe modern capitalism as finance capitalism; because its chief master and its main organizing and guiding force is banking capital which provides the resources for the industrial and trade turnover. If co-operative credit develops and contributes to a powerful inflow of resources into the peasant economy, finance capital will also acquire a guiding and all-determining role in the countryside. But in this case the capital will be public capital, owned by the population itself.

The considerations set out above give an entirely new significance to the modest work of our co-operative associations. They mean that this work, despite its everyday character, is of the very greatest importance for the creation of a new social and economic system. The entire system of agricultural credit, from the local association to the Central Agricultural Bank of the USSR, are therefore of exceptional importance for the building of socialism in our country. However, they give us no indication as to the financial rules by which a credit co-operative must be guided when resolving the problems which it faces with regard to the organization of small-scale credit.

When establishing the foundations of the financial policy of a credit co-operative, account has to be taken of a number of diverse economic elements. The most important of these is the nature of the need for resources, the demand for capital which is presented by the local peasant economy. Depending on regional differences in the structure of the national economy, this will significantly change both qualitatively and quantitatively.

It is continuous and very acute in regions undergoing a transition from an outdated agricultural system to one which is more advanced and more capital-intensive. The acquisition of complex agricultural tools, of better breeds of cattle, of fertilizers and of other ancillaries of an intensive economy require a significant increase in the household's basic and turnover capital, which is achieved by the obtaining of credit.

An entirely different type of need for credit exists in regions whose organization is relatively static. In this case, most of the demand for resources arises from the shortage of turnover resources, from the need to replace the depreciating elements of basic capital and, in part, from the expansion of capital caused by the growth of peasant families and the corresponding expansion in the volume of their economic activity. The need for credit in regions of this type is to a considerable extent seasonal. It must also be noted that the extent of the need for credit does not remain identical from one year to another, but changes in accordance with the evolution of local economic life. The need for resources can itself be broken down into a number of categories, each of which has its own specific character. At the same time, the scale of the need for credit is to a large extent determined by credit conditions. Oppressive, usurious credit reduces the volume of the credit turnover, while any alleviation of credit conditions considerably expands this turnover, making it possible for money to be borrowed to launch economic enterprises which would not be advantageous with high interest rates on loans.

A familiarity with a region's need for credit, with the quantitative extent of this need, its seasonal expansion and its dependence on the level of interest rates – all this constitutes the most important initial information when organizing small-scale credit. When the extent of the need for credit has been ascertained, an active policy is developed so as to find the resources necessary to satisfy it. A number of ways of raising these resources are known in the practice of credit operations.

Given the conditions of the modern Russian countryside with its extremely underdeveloped capital accumulation, and given the population's distrustful attitude towards co-operatives, the latter are, from the financial point of view, the best apparatus for the distribution of state loans. However, this state of affairs should be regarded as temporary and, as the well-being of the peasantry improves, it is to be expected that co-operatives will be able to attract other kinds of resources for their work.

The basic problem with organizing co-operative credit – and indeed with all organized credit – lies in the organization of intermediary links between individuals who need capital in order to run their farms and those who possess unused or relatively unused capital and are able, in one form or another, to make this available for use by those in the first group.

In this respect, credit is usually supposed to represent, as it were, a special kind of trade, where the commodity is the right to use capital, measured by the amount of the capital and the duration of the period of use; and the price for this commodity is the rate of interest payable by the borrower to the lender.

Because credit transactions are of a unique character and are entered into directly between the owner of the capital and its ultimate user, the price of the credit granted is usually fortuitous; and the credit turnover does not of itself give rise to a capital market. However, as soon as credit transactions develop on a mass scale, credit intermediaries will appear on the economic scene; capital circulating in the credit turnover begins to become depersonalized, the price of credit begins to become homogeneous under the pressure of manifest competition in the supply of and demand for capital and there arises a new economic phenomenon: a capital market with its own structure and its own laws of life, similar in some respects to the commodity market. In many ways, private credit transactions will also become subordinate to the pressure of the laws of this market.

During its historical development, the capital market gradually evolved special types of apparatus which served to organize the trade turnover. One such type of apparatus, which serves local markets for peasant households within the boundaries of the sub-district [*volost'*], is the credit association.

The basic task of the credit association is to organize the local

Co-operative Credit Societies 87

Figure 4: Normal development of the credit turnover

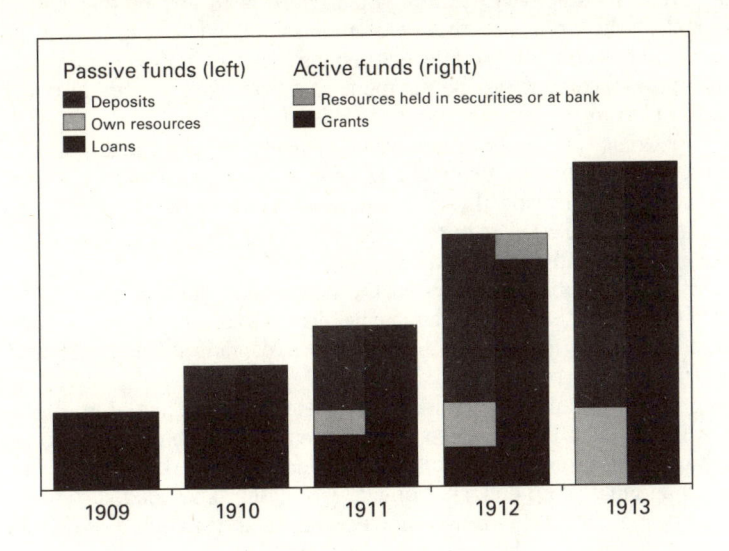

Passive funds (left)
- Deposits
- Own resources
- Loans

Active funds (right)
- Resources held in securities or at bank
- Grants

1909　1910　1911　1912　1913

Figure 5: Abnormal development of the credit turnover

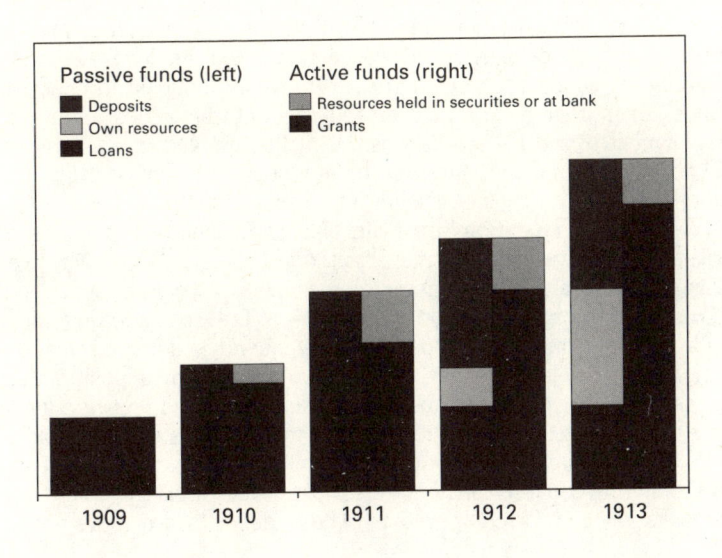

Passive funds (left)
- Deposits
- Own resources
- Loans

Active funds (right)
- Resources held in securities or at bank
- Grants

1909　1910　1911　1912　1913

capital market for fairly large groups of households which comprise the area of its activity. The first task here is to attract the unused capital in the area into the association's bank. This capital will be provided by members of different economic strata in the countryside; and the inducement will be payment of a price for the capital in the form of a rate of interest on money deposited.

A reading of co-operative balance-sheets – and these make a fascinating subject for economic reading – gives us nowadays a vivid picture of healthy and thriving co-operative work, of its growth and decline, as well as a picture of the malfunctioning of co-operatives and of their gradual fading away.

Figure 4 graphically shows us the normal and healthy growth of a credit co-operative, which began by productively deploying a state grant and which gradually succeeded in expanding its deposits thus relying on local resources.

Another example of an entirely different kind is depicted in Figure 5, which shows how a credit co-operative was unable, owing to its poor organization, successfully to develop the work which it had begun in building up deposits; how it was all the time obliged to make use of capital attracted from outside, while at the same time failing to deploy the capital which it had assembled.

We can also see, when observing the development of balance-sheets, the manifestation of an active financial policy of co-operative credit (see Figure 6).

Thus we can see, for example, how under the pressure of an intensified need for money and under the pressure of requests for loans, a credit association was obliged in 1909 to raise the rates of interest on its deposits, and consequently on its loans as well, thereby attracting resources. However, having doubled its 'passive funds' over four years and in the face of difficulties over the deployment of the resources which it had assembled, it was obliged to reduce the rate of interest on its deposits, thus putting the composition of its balance-sheets into a more healthy state.

We see a very curious example of financial policy in a series of balance-sheets, which we have depicted in Figure 7. A co-operative, having consolidated its position after a number of years of successful work, tried to organize long-term credit – at first, so as to dovetail it with all the long-term deposits. Subsequently, it relied on the constancy of deposit funds including the short-term deposits, thus breaking the basic rule of banking according to which the active funds of a credit centre must be totally determined by its passive funds.

No private bank wishing to avoid bankruptcy can possibly tie up more of its resources in long-term loans than it has in long-term deposits. But for a co-operative, whose deposits are more stable

Co-operative Credit Societies 89

Figure 6: Regulation of the credit turnover by changing the level of funds (percentages)

Passive funds (left)
- Loans
- Deposits
- Long-term deposits
- Own resources

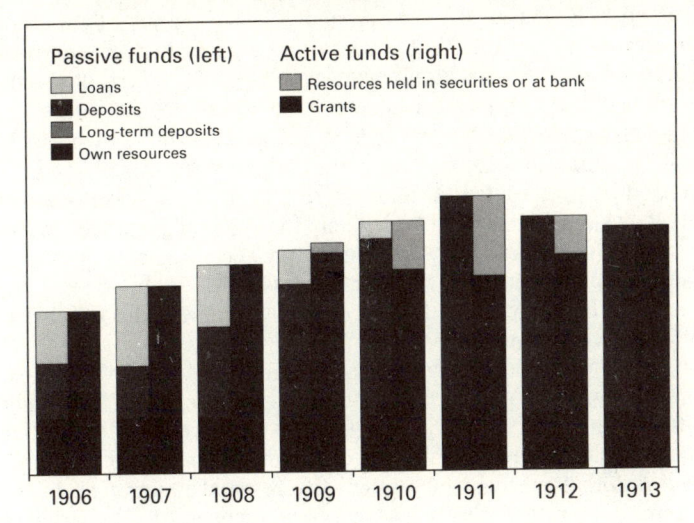

Active funds (right)
- Resources held in securities or at bank
- Grants

1906 1907 1908 1909 1910 1911 1912 1913

Figure 7: Introduction of long-term credit

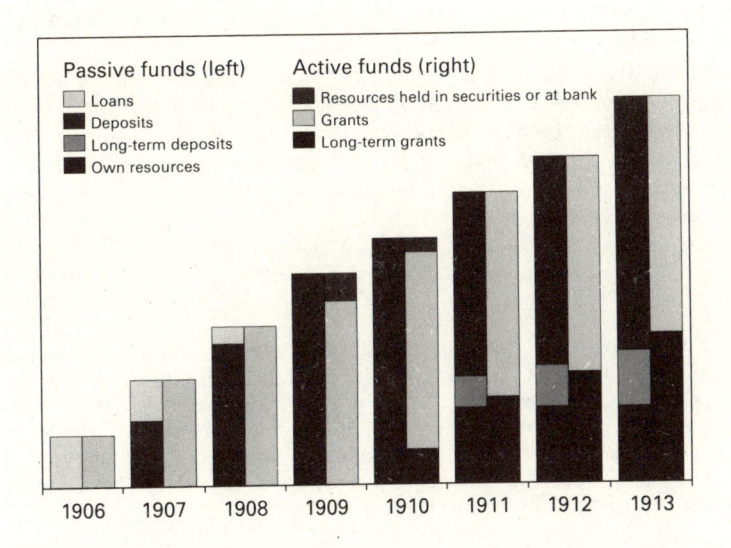

Passive funds (left)
- Loans
- Deposits
- Long-term deposits
- Own resources

Active funds (right)
- Resources held in securities or at bank
- Grants
- Long-term grants

1906 1907 1908 1909 1910 1911 1912 1913

even though they are in a formal sense short-term, it is, to a certain degree, entirely possible to break this rule.

It is entirely conceivable that an active policy pursued by a credit co-operative may extend not only to the organization of long-term credit, but also to the encouragement, with the aid of credit, of specific new branches of the peasant economy, for example, dairy farming.

The association's profits are used in order to create its own special capital for this purpose; and with the help of this capital credit is organized on preferential terms so as to encourage the acquisition of milch cows and the purchasing of the essential fodder. This preferential credit can also be used to finance not only individual households but agricultural co-operative associations. In general, the financing of other types of co-operation is an essential obligation of local credit co-operatives.

In addition to the examples just examined relating to financial policy for the development of a credit co-operative, its tasks include the management of its own balance-sheet in accordance with the state of the peasant money economy during different seasons of the year.

In order that we may fully grasp the national importance of the economic features of the seasonal policy of peasant co-operatives shown in Figure 7, we would direct the reader's attention to a comparison between the money turnover of the co-operative apparatus and the money turnover of the peasant economy.

A considered examination will convince us that credit co-operation is a form of gradual socialization of a very large part of agricultural capital – which is now more fully and more appropriately used for the development in the public interest of the productive potential of the countryside.

The harnessing of this capital for consciously formulated social purposes endows the organized public co-operative apparatus with an exceptional degree of power over the development of the agricultural economy, and enables it, through its financial influence on peasant households, to lead them along the path of the advancement of agricultural production.

NOTE

1. Editor's note: The 'Principles of Raifeizen' were broadly adopted by the credit societies of the European Co-operative Movement during the period in question. They included full mutual responsibility of the Societies' members towards each other, the restriction of loans to productive investment and to Societies' members, as well as the assumption that each Society would be small and be led by unwaged officers elected by the membership.

5

The Peasant Family's Money Economy and its Organization on Co-operative Principles

The disintegration of the system of the natural economy in the countryside, and the gradual involvement of the peasant household in the commodity turnover, has more than once been the subject of extensive investigations. However, the literature has offered hardly any study at all of the organizational changes caused by the development of money transactions in the household's internal structure.

We know practically nothing about the particularities of this revolution. We have no clear notion as to what changes the peasant household has to make in its organizational plan in order to make the transition from a natural type of economy to forms of commodity production. But these changes are very significant, and only after becoming familiar with them can we fully understand the nature of the money economy which is operated by the peasant homestead.

Let us try, as thoroughly as possible, to grasp the organizational patterns of both types of household.

A natural economy is the most vivid example of an economy of the consumerist type. Within it, all the individual items of the consumer budget have to be satisfied from its own output. As a rule, these items are diverse and numerous; and in order to cater for this variety, the peasant family had to develop a no less complex and varied plan of production. A household that aims at satisfying in kind scores of family needs must naturally have scores of different sectors. Therefore a *natural* economy was always the most complex of all modes of agricultural production.

It is true that at the present time such a form of natural economy is very hard to find. The structure of the economy has, almost everywhere, been penetrated by elements of money relationships, which have simplified its organization. The development of the market made it possible and profitable to abandon many small sectors and, by expanding the production of basic and more profitable goods, to dispose of the surplus portion of the harvest on the market. The money obtained from these sales financed the items of the consumer budget which, in a natural economy, had to be catered for through the organization of special sectors of production.

A simple comparison between two peasant households – one of which is organized on principles close to those of the natural system, and the other of which is based on commodity production – provides clear confirmation of what has just been said. Let us take the *Zemstvo*'s book of statistics for the Totem district of the Vologda province which is one of the remotest corners of our country and one of those most heavily based on the natural economy; and let us excerpt the final columns of the budget tables, relating to the groups of households which were largest in scale. Let us then compare the average figures arrived at with those for the ordinary household in Volokolamsk.

Table 16 makes it possible, at first sight, to identify substantial differences. We can see that money expenditure as a percentage of the overall amount of the consumer budget is, in the case of the households in Totem, equal to a total of 20.9 per cent; whereas in a household in Volokolamsk it reaches 66.3 per cent. In other words, the peasant household in Totem represents an economic organization which is to a certain degree isolated and has few social or economic ties with the outside world. By contrast, the peasant household in Volokolamsk has already become drawn into the world's economic turnover; and it lives not only by its own income but by a share of the common national income. An economy structured in this way was naturally bound to affect the manner in which its production was organized.

The numerous items of the consumer budget, which in the Totem district are satisfied in kind, make it necessary for the household to have a complex organization, providing 32 different kinds of product. However, in the Volokolamsk district, only ten items of the budget are satisfied in kind, which makes it possible greatly to simplify the structure of its economic organization.

The comparative complexity of the economic structures of the two households in question can be partly judged from Table 17. Thus in the Totem household, 87 per cent of all income is consumed in kind within the household; and its production is determined by the needs of consumption. However, in the Volokolamsk household, only 39.6 per cent of what is produced is intended for direct consumption by

Table 16 Structure of the consumer budgets of households based on a natural economy and households which are market-oriented

(Expenditure in money and in kind in roubles)

Amount consumed:	Totem district		Volokolamsk district	
	In kind	In money	In kind	In money
Rye	58.5	–	26.0	40.0
Barley	13.3	–	–	
Wheat	9.5	0.6	–	7.5
Oats	4.4	0.1	5.0	–
Malt	3.9	–	–	–
Groats	7.8	–	–	13.5
Peas	3.8	–	–	–
Potatoes	5.8	–	12.0	–
Cabbages	0.3	0.0	–	–
Cucumbers	0.1	–	–	11.0
Onions	1.3	0.0	1.0	–
Other vegetables	1.7	–	–	–
Vegetable oil	2.2	1.0	18.8	–
Mushrooms	4.1	–	–	–
Berries	2.3	–	–	–
Payments for pig feed	–	3.6	–	4.5
Beef	3.9	1.2	–	6.0
Veal	1.8	–	20.0	–
Mutton	3.9	–	–	–
Pork	6.8	1.4	–	5.1
Eggs	5.2	0.0	0.5	–
Milk and dairy products	51.3	–	150.0	–
Poultry	0.2	–	0.5	–
Fish	2.1	4.5	–	10.0
Salt	–	1.8	–	2.0
Seasoning	–	0.6	–	16.8
Tea and sugar	–	11.8	–	50.0
Tobacco	–	0.3	–	–
Alcohol	3.5	6.1	–	21.0
Hops	0.1	0.5	–	–
Clothing	4.3	10.8	–	145.0
Cards	0.0	0.3	–	3.0

Table 16 *continued*

	(Expenditure in money and in kind in roubles)			
	Totem district		Volokolamsk district	
Amount consumed:	In kind	In money	In kind	In money
Spinning of flax	4.0	–	–	–
Wool	2.5	–	–	–
Sheepskin	1.2	–	–	–
Soap	–	1.1	–	12.0
Lighting	–	1.9	–	4.0
Firewood	8.6	3.6	20.0	50.0
Utensils	0.0	1.8	–	2.0
Spiritual needs	–	4.8	–	4.5
Total	218.4	57.8	253.0	497.8
	276.2		750.8	
in percentages	79.1	20.9	33.7	66.3
	100.0		100.0	

the family (see Table 17). The remaining 60.4 per cent is offered for sale on the market and it provides for the family's consumption only in the sense of making it possible to acquire goods which are needed out of the proceeds of sale.

Households in other areas, for which budget data exist, show differing ratios of commodity output in relation to total output; and these households occupy an intermediate position between the extreme types examined above. Thus budgets for the Vologda province show a rate of dependence on money relationships of 34.6 per cent for the Vologda district; while for the Vel'sk district they show a corresponding rate of 10.7 per cent. An investigation of budgets in Starobelsk showed a rate of 21.1 per cent. For the Smolensk province the results which we obtained were from 24.3 per cent to 11.1 per cent.

We know of accounts of peasant households in other countries which have shown a higher degree of commodity output. Thus, the description given by E. Laur of Swiss households showed a rate of dependence on money relationships of 62 per cent. And we know of other monographs on the peasantry in the West, which give very

The Peasant Family's Money Economy and its Organization 95

Table 17 *Receipts in money and in kind of households in Totem and Volokolamsk*

	Totem district		Volokolamsk district	
	Consumed in kind (roubles)	Sold to the amount of (roubles)	Consumed in kind (roubles)	Sold to the amount of (roubles)
Rye	74.4	6.5	27.0	–
Barley	21.7	–	–	–
Wheat	12.5	0.7	–	–
Oats	59.5	19.4	55.0	–
Potatoes	7.5	–	18.0	–
Flax seed	2.1	0.8	25.0	140.0
Flax fibre	5.6	3.3	–	306.0
Peas	4.3	–	–	–
Cabbages	0.3	–	–	–
Cucumbers	0.1	–	–	–
Onions	1.2	–	1.0	–
Other vegetables	1.7	–	–	–
Beef	4.0	–	–	–
Veal	2.1	–	20.0	–
Mutton	3.9	–	–	–
Pork	6.8	–	–	–
Milk and dairy products	52.1	7.6	150.0	–
Hides and wool	5.8	0.5	1.0	7.5
Poultry products	0.6	0.6	1.0	–
Total	266.1	39.4	298.0	453.5
Receipts from crafts and trades	–	48.9	–	85.0

similar figures. We know of no case where the rate of dependence on money relationships in peasant households exceeded 70 per cent. This evidently represented an upper limit since, owing to the very nature of agricultural production, even on purely capitalist farms, some of its products will be consumed within the farm, feeding the workers or the proprietor's family and providing fodder for cattle and seed for sowing. These products would not be replaced by those which have been bought because their cost within the household is considerably below the purchasing prices on the free market.

Households high up on the monetary scale differ from those based on a natural economy not only because their organizational plan is a good deal simpler, but also because of substantial differences in the way they make their economic calculations.

In the case of a household based on a natural economy, the activity of the person carrying on its economic management was directed towards satisfying particular consumer requirements; and to a great extent it had a *qualitative* emphasis. The family had to be provided with specific products – precisely those and no other. *Quantity* could be measured only in relation to the extent of each particular need which was being catered for: by reference to whether 'there was enough' or 'not enough' and in the latter case how acute was the shortage. And because of the elasticity of the needs themselves, there could be no great precision about any such measurement.

Therefore, in a household based on a natural economy, the question could not arise as to whether, for example, it was more profitable to sow rye or to mow hay; since these activities were not interchangeable and therefore lacked any common yardstick for comparison. The importance of the hay produced was measured by reference to the need for fodder, and the importance of rye by reference to the feeding of the family. It might even be asserted that the poorer the quality of the hay, and the greater the effort needed to produce a given quantity of hay, the greater was the importance attached to hay-making.

But as soon as a household becomes part of the commodity-money turnover, its aims assume an entirely different character. Economic activity abandons its qualitative emphasis. Needs can now be satisfied by means of products which have been bought. What comes into the foreground is the 'quantitative' interest – in *producing the greatest possible quantity*, which, once it has been exchanged, can assume whatever qualitative form is necessary for the purpose of buying goods. As the dependence on money relations progressively develops, the 'quantity' obtained becomes increasingly separated from 'quality'; and it begins to assume the abstract character of a value.

Once the exchange of commodities has become widely developed, it becomes a matter of economic indifference to a household as to where it should concentrate its labour – provided only that its labour is fully utilized and is well rewarded by the market value of what it has produced. And since the level of remuneration for the labour invested in producing different goods is ultimately determined by the state of the market, it is therefore self-evident that, as commodity elements progressively develop within the economy, the organization of a household (which, in the context of a natural economy, is determined in all its details by the family's consumer requirements) will increasingly become subordinate to the influence of the market situation.

A fall in the price of any particular product will lead those engaged in farming to stop producing it and to sow the fields with another crop. If the price of milk improves in relation to the price of bran, this immediately leads to an expansion of milk production, and so forth. A household that has 'learned the meaning of weights and measures' will begin to play the market just like a dealer on the stock exchange.

Such is the basic organizational and economic significance of the transition from a household based on a natural economy to one based on commodity-money relations. The household becomes freed from the 'qualitative' influence of the requirements of consumption and by turning to the production of commodities, which are constantly adapted to the changing situation on the market, it gains the opportunity of acquiring a significantly greater number of valuable things and, consequently, of increasing the remuneration for its labour. However, this rise in income makes it necessary for the household – quite apart from the organization of its production – to carry out a further, complex, organization of its money economy: in other words, it has to organize its relationship with the market.

It must immediately be noted, however, that the money economy of a peasant homestead is in many ways the antithesis of the trading operations of a commercial firm. Trade has been described as 'the buying of commodities for the purpose of their subsequent resale'. Its operations centre on the difference between purchasing and reselling prices: therefore the absolute level of prices is of no importance to a trading organization. But a peasant household, like any other enterprise which produces raw material, has to organize the sale of the products of its labour in order to buy, out of its money receipts, its essential means of livelihood and its implements of production. A peasant household is therefore interested in the absolute level of prices of agricultural products; or it is, at all events, interested in a general rise in the market prices of agricultural commodities as a whole, relative to those of the industrial commodities which it acquires.

In the same way as the skilful and successful organization of a trading operation can lead to a considerable increase in profits, so the skilful organization of the peasant money economy – that is, its purchases and sales – can be very advantageous to the family carrying on the economic management of a household and can enhance its level of well-being. The art of buying a commodity of an appropriate quality at the right time and at the right price or of selling one's products to a substantial purchaser who can take them without cheating over weights and can pay a 'real' price corresponding to the quality of the goods – this art is of great value to a family farm. The proper and skilful organization of purchases and marketing is no less beneficial to such a farm than is the skilful management of agricultural production itself.

Before making any generalizations on this subject, we think it essential to familiarize ourselves empirically with the ways in which different kinds of peasant household organize their money turnover.

The Budget Studies of peasant households, which have been carried out in many different parts of Russia, have shown great diversities in the structure of the money economy. In order to reduce all types of receipts and expenditure to a common denominator, we shall first of all divide them into those that are *natural* – that is, obtained from the household itself and consumed within the household or expended on the needs of production; and those that take the form of money, that is, receipts from selling one's own labour or the products of one's labour, and expenditure involving the payment of money in order to acquire goods; or expenditure on production or personal requirements.

We can see, by looking through Table 18, that the money budget of the families of middle peasants fluctuates around the level of approximately 200–500 roubles within an overall budget of 700–1,000 roubles; so that 25–50 per cent of the budget is based on money transactions. In other words, the income and level of well-being of a modern peasant family are dependent to the extent of one half of its ability to organize its money economy. If the peasant is able, as a result of this skill, to obtain an increase of 10 per cent in the price at which he sells the products of his labour and can use the money so earned in order to buy 10 per cent more of the goods required for the everyday needs of the farm and of the family, then his trading skill will have brought him an overall increase in the level of his well-being equal to one-fifth of his money budget or about one-tenth of his overall budget. And this increase can often be of decisive importance in making it possible to renew or expand the peasant household's capital.

Table 18 Receipts and expenditure of peasant households in money and in kind

Name of district	Receipts				Expenditure			
	In kind	In money	Total	Percentage based on money transactions	In kind	In money	Total	Percentage based on money transactions
Volokolamsk	670.0	528.1	1198.1	44.2	554.9	500.1	1055.0	47.3
Gzhatsk	451.9	247.0	713.9	34.4	463.2	251.1	714.3	35.2
Porech	621.0	198.6	819.7	24.2	628.0	198.6	826.7	24.2
Sychev	485.5	288.2	773.7	37.3	488.0	284.1	767.1	37.0
Dorogobuzh	650.1	180.3	830.4	21.7	640.2	213.4	853.6	25.0
Starobelsk	568.1	442.0	1010.1	43.7	499.0	436.5	934.5	47.7
Vologda	238.5	209.6	548.1	38.3	238.7	217.7	556.4	39.1
Vel'sk	361.2	121.9	438.1	27.8	317.0	123.5	440.5	28.0

100 *The Peasant Family's Money Economy and its Organization*

The drawing-up of the money budget, as we shall see below, is a far from easy problem for the peasant household and, once again, it has been little studied. Our own budget studies indicate that the structure of this budget may be of very various types.

In most Russian peasant households money receipts from the sale of agricultural products are supplemented by receipts from the sale, outside the household, of the labour which for some reason cannot be utilized within it. Earnings derived from trades play an outstandingly important role in the Russian peasant household. Market conditions, as well as the local economic situation, will sometimes increase and sometimes reduce the extent to which individual sectors of the economy are oriented towards the market;

Table 19 *Total money incomes and money incomes from trade for an average peasant household (in roubles)*

Provinces and Districts	Total money income per household	Including individual employment in crafts and trades	Income from trade and and craft enterprises	Total income from crafts and trades in percentages
Vologda	229.1	89.1	5.9	41.5
Novgorod	252.5	86.7	24.3	43.6
Smolensk	301.6	99.8	9.2	36.2
Kostroma	263.5	140.1	9.5	56.8
Vladimir	515.9	295.0	–	57.1
Moscow	704.8	302.4	36.8	48.1
Tula	256.1	111.5	18.6	50.8
Ryazan	289.2	133.3	23.4	54.3
Voronezh	169.9	41.9	0.1	24.7
Penze	152.7	40.6	4.3	29.7
Tatar Republic	148.8	57.4	11.6	46.3
Saratov	224.3	33.6	3.1	16.4
Armavir	439.2	56.8	1.6	13.3
Novonikolayev	168.8	45.6		27.0
Barnaul	218.5	44.6	27.0	32.9
Chernigov	160.7	34.1	28.1	38.5
Berdichev	152.5	26.0	–	17.1
Poltava	198.5	26.6	38.6	32.9
Starobelsk	196.8	59.0	18.7	39.4
Turkmenistan	433.7	49.1	13.3	14.3

the main source of money receipts will sometimes be agriculture and at other times cattle-rearing. But income derived from trade nearly always represents an impressive proportion of money receipts.

Even now, when crafts and work away from the village have by no means been fully restored, money receipts connected with them nevertheless occupy a very prominent place in our peasants' budgets. Thus, for example, a budget investigation carried out by the Central Statistical Board (TsSU) in 1923–4 provided the data shown in Table 19.

We can see how varied is the structure of the money budget, depending on the economic situation which surrounds the household. The household seeks to rely for its money turnover sometimes on the sale of flax and at other times on disposing of its cattle or its grain. But most commonly of all, unfortunately, it relies on hiring out its own labour for the purpose of earning money. Despite the dysfunctional nature of this method of solving the problem of organizing the money budget, it is widely encountered and can take extreme forms.

Agricultural households, despite the great importance of money transactions in their consumer budgets, nevertheless remain to an exceptional degree based on a natural economy. More than that: agricultural output not only fails to provide cash, but the maintenance of this output requires considerable expenditure out of earnings derived from non-agricultural trades.

It is scarcely necessary to point out what hardship such a system causes – not only to the economic life, but also to the social life of our countryside. Adult men are away from their families for periods of five to six months or even longer, leaving the farm to be run by women who are not physically strong enough to do so adequately. The Russian peasant woman ploughs the fields and does the mowing as well as threshing the grain. There are a good many families in which the men have grown utterly unaccustomed to agricultural work and who, as their wives say, 'do not even know how to harness a horse'. A 'women's household' which is feeble has little productive potential and is of little value from the point of view of the national economy.

There is no doubt that one of the main factors which gave rise to this type of structure, especially in industrial areas and areas of agrarian overpopulation, was the shortage of arable land. This shortage prevented the peasant family from making ends meet within the existing system of farming and compelled it to send its surplus manpower into non-agricultural trades.

However, even if the family had been able, through strenuous effort, to earn an income from agriculture which was sufficient for a livelihood, nevertheless – given the pre-war level of prices for agricultural products – the peasants would have had many reasons for abandoning agriculture. The most important of these reasons was the higher remuneration of labour outside agriculture. Once a peasant labourer got the opportunity to earn 1½–2 roubles per working day outside agriculture, he naturally *'had no time'* for agriculture which paid him only 70 to 80 kopecks for the same day's work.

These have been the two most important factors which lead to the break-up of the agricultural way of life in our countryside, which drive its population into seasonal work and flood our cities with cheap, semi-skilled manpower and an army of unemployed; and which give to our urban working class a character which is half-proletarian and half-peasant. The resulting situation can be eliminated only when agricultural production becomes the most advantageous of all possible occupations for a peasant family.

It is obvious that such a return by the population to the land will be possible only when:

1. The relative price level for agricultural products rises to the point where the remuneration of labour employed in agriculture is higher than its remuneration in other occupations; and
2. The organization of agricultural output is rationalized and intensified to such a degree that the existing peasant allotments are able to earn incomes sufficient to meet the family's consumer budget.

The first of these conditions depends to a considerable extent on the situation of the world market. But we still have to recognize that, where the market situation remains constant, the level of selling prices depends to a great extent on the seller's ability to organize his sales and on the way that the money economy is managed. It is posssible to sell one and the same *berkovetz* [approx. 3.21 hundredweight] of flax in one and the same year either for 40 or for 50 roubles, depending on the skill of the seller.

How then can this skill manifest itself? How is the peasant able – by improving the organization of his purchasing and marketing – to enhance the level of his well-being? To answer this question we have to assess the market relationships which link the peasant households with the world market. The improvement of these relationships is the purpose underlying the organization of the peasant family's money economy.

An attentive observer who follows the journey of a bag of flax or a sack of wheat sold by a peasant can usually discern a highly complex chain of social and economic interrelationships which arise as the commodity proceeds on its journey. The historical development of markets gave rise to a whole system of complex trading mechanisms for the purpose of conveying commodities from the producer to the consumer. These mechanisms evolved – in ways which depended on the nature of the commodity – into a whole series of consecutive links.

There is the small buyer and cattle-dealer, engaged in buying up commodities in the villages and at the bazaar. There is the local trader who already has his own warehouses for processing the commodities and grading them. There is the whole network of brokers, working for commissions or working on the stock exchange. There are the large wholesale firms and the exporters and importers. Alongside them are the ancillary enterprises related to transport, insurance and banking, which finance trade. There are the companies concerned with the investment of capital. And, finally, there are those who just speculate on the commodity stock exchanges, who trade on the rise and fall in prices. These are the specific agencies which, in their mutual rivalry, perform the complex role of conveying agricultural commodities from the producer to the consumer in a capitalist market.

The balance of economic factors, which is arrived at after many years of fluctuation, creates within each of the markets listed above a system of prices for all primary, intermediate and finished products, which ensures that every agency in the market receives an adequate profit from its work and that it therefore has a direct economic interest in doing the work in question. The representatives of commercial capital are able – through their skill and commercial adroitness – to obtain profits in excess of the level of normal profits, established by the correlation of prices; and this superprofit is usually obtained at the expense of the producers and the ultimate consumers, who are scattered and unorganized.

Such is the picture of market relationships in which the peasant family has to build its money economy; and such is the structure of market trading, created and consolidated by the practice of centuries. It is powerful owing to its degree of organization and its technical experience; and it has every incentive to obtain from the peasant the products of his labour at the lowest possible price; and to provide him with the means of production and consumption at the highest possible price.

The peasant household drawing up its money budget is confronted with the relentless pressure of powerful capitalist organizations which get their profits by underpaying for the products of peasant labour and overcharging for the commodities which the peasants buy.

We are faced with the usual picture of the peasant masses being utterly in the grip of commercial capital and of a social and economic struggle to protect the remuneration for peasant labour. When defending its 'wages', the peasantry needs to strengthen its position in every way in order to obtain on the market the highest possible remuneration for the labour which it has invested in producing agricultural products, which it eventually exchanges for commodities bought on the market.

In order to strengthen his position in this bitter struggle the peasant has to try to ensure:

1. That his commodity is sold at a time of year when the state of the market is most favourable to the seller; that is, when its supply is limited, when it is acutely in demand and when its price is high. And conversely, the peasant must buy the commodities which he needs at times when they are in the most plentiful supply.
2. That the commodity being offered on the market is graded in a manner appropriate to its quality; and that it is packed and offered for sale in the kind of packaging which will lead the market to pay the full value of its quality. The correct valuation of a commodity by its purchaser may not uncommonly result in the virtual doubling of the producer's income from his labour.
3. That a commodity which may be subjected to primary reprocessing is offered to the market in the kind of reprocessed condition which will stimulate the greatest demand and result in the greatest remuneration for the labour invested in it.
4. That the household should, as far as possible, offer the commodity to the part of the market which is situated nearest to the ultimate consumer. Apart from getting fairer conditions for the acceptance of the goods in such a market, one can also expect higher prices, since in this case the peasant can by-pass the middlemen and take for himself the middlemen's profit which has been established by the balance of market factors. In just the same way, a farmer, when he buys goods, must try whenever possible to buy them at first hand in order to get a wider range of choice and lower prices.
5. Lastly, that the household should be flexible in its output, always capable of adapting the type and grading of the goods which it produces and capable of responding to constantly changing market requirements.

Such are the difficult tasks which face the peasant homestead when it organizes its money economy. A small, economically feeble peasant household needs to display exceptional energy, sense and skill in order successfully to solve even some of the tasks which we have set out. Some of them may be basically insoluble for a small-scale household.

For peasant households, therefore, there is only one reliable way out of the situation, and it begins to assume an exceptional importance. It lies in the possibility – through the organization on co-operative principles of many thousands of households – of enabling the peasants to create their own powerful, specialized organizations, which organize the peasants' money budgets by setting up their own large-scale trading apparatuses, which serve the peasants and are managed by the peasants. In this case, the peasants are resisting capitalist exploitation with its own weapons: powerful enterprises, large-scale turnovers and perfected techniques.

These powerful collective organizations are able, by attracting into their turnover resources from credit institutions, to carry out buying and selling operations at the times most favourable to the peasant economy. They can provide a commodity with the kind of grading and external appearance which no individual peasant would be able to do. In just the same way, by setting up butter manufacturing, potato-grinding, vegetable-drying and other co-operative factories, they can offer a commodity to the market in the kind of reprocessed condition which is most advantageous to the seller. And because this reprocessing is mechanized, it is considerably cheaper for the peasant than is reprocessing carried out in domestic conditions. It is unnecessary to add that co-operatives have undertaken buying and selling operations worth millions of roubles and have worked in the very largest wholesale markets. Therefore, they have been able to buy and sell at the most advantageous prices – and hand over to the peasantry the whole of the middlemen's profit.

Finally, the good knowledge of the market which co-operative centres naturally possess, as well as their ability to enlist the help of agronomists and technical specialists, make it possible for co-operatives to become a powerful factor which influences the internal organization of the economy and restructures the economy so as to make it conform more closely with market conditions.

Such is the exceptionally important assistance which co-operative principles can offer in the organization of the working peasant family's money economy. The ideas set out above are clear and are, at first sight, extremely easy to implement. However, it was only after nearly a century of organizational inquiries as well as thousands of distressing bankruptcies, that it became possible to hammer out the organizational principles to bring us close to solving the tasks set out.

But how should the member of a co-operative staff begin his

organizational work? To what, above all, should he devote his attention? By what basic principles shoud he be guided in his struggle for markets? And to what should he devote his particular attention when studying the market from a co-operative point of view?

It must be noted first of all that when a member of a co-operative staff sets out to achieve co-operative intervention in the organization of the marketing of agricultural products, he must begin by considering in what order his work should proceed.

Peasant households in all parts of Russia produce and sell a whole number of varied agricultural products. There is no doubt that the ultimate aim of the rural co-operative movement is to organize the marketing of all these products on co-operative principles. But at the present time we have to confine ourselves to a task which is within our power; and we must direct all our efforts towards solving it, without wasting our effort on other endeavours which we know to be futile.

We have to select two or three products which, because of the way that their markets are organized, can most easily be organized on co-operative principles, and which, at the same time, are of major importance for the national economy. It is on them that we have to concentrate our co-operative effort. Only if the work is organized in this way do we have any guarantee of success or are we able to gain the organizational experience needed for organizing the more difficult markets on co-operative principles at a later stage.

But on what principles should we choose the markets for inclusion within the co-operative system in the first instance? We know that the methods of co-operative research into markets have up till now by no means been adequately developed; and it is therefore extremely difficult to give a complete answer to this question.

However, the practice of co-operative work does make it possible to identify certain fundamental stages which have to be passed through when we seek to ascertain how easily a market can be organized on co-operative principles. While acknowledging that marketing co-operatives will, at the initial stage of their work, always be short both of resources and of technical trading experience, we must nevertheless note those market conditions which are conducive to a co-operative's success.

Thus, if we turn our attention to trading technique, we can very confidently say that the greater the homogeneity of a product and the more it is susceptible to depersonalization, the easier it will be to organize the product on co-operative principles.

The more stable the quality of the product and the greater its homogeneity, the easier it is to conduct trading operations in the product. This is a matter of particular importance for those co-operative staff workers who still lack experience in matters of trade. If the product in question is not absolutely homogeneous, but represents a whole range of varieties and grades, then we can, for the same reasons and with equal confidence, assume that the organization of the marketing of this product on co-operative principles will be all the easier, the more stable its grades and the more easily it can be made subject to classification and selection. If the grading of the product is easy and if it is subject to firm standardization, this will be a guarantee of success for co-operation.

It is enough to give two or three examples, in order to explain the propositions just put forward. Thus, for example, we assume that the marketing of hens' eggs can easily be organized on co-operative principles because the collection and grading of this product do not require very much technical equipment and those who work in co-operatives will very soon be able to bring the grading system up to the technical level established by commercial capital.

Conversely, the inclusion of flax in the co-operative system presents extraordinary difficulties; since a product which comes from one and the same farm is often exceedingly diverse; its grading is made more difficult by the absence of any precisely established standards and by the difficulty of precisely ascertaining the quality of the fibre without complicated adaptations. The grades of flax will vary over small areas; and one and the same grading description may, in different years, refer to goods of different quality. It can therefore be confidently assumed that the organization of flax on co-operative principles will get into its stride only after the question has been satisfactorily resolved of how to organize the grading.

If to these prerequisites for the success of co-operative marketing we then add the prerequisites for the safe preservation of the product – and if we say that the organization of the marketing of a product on co-operative principles will be easier, the less the product is vulnerable to damage – then we shall have outlined the whole range of technical conditions which the co-operative staff worker has to take into account.

It follows from this last rule that a co-operative staff worker who lacks sufficient experience will have difficulty in coping with such refined commodities as flowers, live poultry, fruits, sucking-pigs and other things that need special care and need to be sold quickly.

We are then faced with a number of economic problems. We must first take account of the general character of the market which we are organizing on co-operative principles. We have to be clear as to whether it is a local market (trading, for example, in milk, hay, vegetables, and so on); or whether it is a market of a regional kind (trading in fruit, cattle, firewood, and so on); or a market of global importance (trading in wheat, butter, eggs, and so on).

The size and breadth of the market is one of the most important preconditions for its organization on co-operative principles. We may assert that the co-operative organization of the marketing of any product whatsoever will be easier, the greater the absorptive capacity of its market. Indeed, in a small market where demand can be fully satisfied extremely rapidly, any fortuitous accumulation of a product will overload the market and lead to a drastic fall in prices and extreme difficulty in disposing of the product.

An example may be provided by the market for fresh milk in a small town. The first dairy partnership in the town finds an adequate market. But as its business develops, it saturates the town with milk with the result that prices fall so that the co-operative has to change over to supplying butter – a product for which there is a wider market. This kind of unevenness in marketing conditions can be handled only by a flexible entrepreneur who is prepared to gamble. But it certainly cannot be handled by a not very experienced co-operative organizer who requires that marketing conditions should be stable.

Professor M. Tugan-Baranovskii, when comparing the success of village co-operatives with the failure of craft co-operatives, had seen this as the direct result of the fact that a successful farmer is supplying an unlimited global market, whereas a craftsman is supplying a narrower, local market.

Apart from the breadth and capacity of the market, a great deal also depends on the degree of flexibility in the consumption of the product. How is this term to be understood? We shall try to clarify it through the following obvious examples.

If we compare the quantities of bread which different people consume, we find that they are very similar to one another. The difference between being underfed and being fully satiated is equal to no more than a rate of 2–3 poods [approximately 72–108 lb] per year. Neither a rise in prices nor a decline in well-being can lead to any drastic change in the consumption of grain products. The levels of bread consumption remain not very flexible. Therefore, the slightest interruption in the supply of grain commodities on the market causes an acute need for bread, a sharp competition between buyers and a rise in prices. Conversely, any surpluses, unless they can be exported to other areas, result in unsold stocks and a sharp drop in prices.

If, as a rule, we do not notice any sharp fluctuations in grain prices, the explanation lies in the extraordinary size of the grain market, as a result of which shortages in one region are made good by surpluses elsewhere. In the case of vegetables, the consumption

The Peasant Family's Money Economy and its Organization   109

of which is also not very elastic, such drastic price changes can be observed very frequently.

We see a different state of affairs when we study, for example, the markets for sugar or cotton cloth, where the rates of consumption are characterized by an extreme flexibility. The consumption of these products when they are in short supply may drop to the barest minimum; and subsequently, when supplies increase, consumption may rise tenfold or more. Such an elasticity in the rates of consumption enables the consumer to be extremely sensitive to any change in price – reducing his consumption at the slightest price rise, while significantly increasing his consumption when prices fall.

One result of this flexibility of consumption is that when large supplies accumulate on the market, it usually needs only a small drop in prices for the consumers to liquidate the surplus by increasing their consumption. It is obvious that the greater the flexibility of the consumption of any particular product, the greater the capacity of the market for that product will prove to be.

Apart from territorial breadth and flexibility of consumption the capacity of the market for any product is greatly affected by the social composition of its consumers.

If a product is consumed only by the prosperous stratum of society (as in the case of expensive fruits, flowers, silk, and so on) then, despite the possible breadth of the market and the flexibility of requirements, the market will continue to have only a small capacity, because the total number of its consumers will be very small. It may therefore confidently be assumed that the greater the extent to which a product is consumed by the population at large, the greater will be the market's absorptive capacity.

Thus, for example, during the years immediately before the war, the Siberian Alliance of butter manufacturing partnerships [*arteli*] was already having some difficulty in marketing butter, because its consumers in Western Europe came from the prosperous strata of the population, whose numbers were limited. However, the middle or poorer classes used coconut oil or margarine. The leaders of the Alliance therefore began to think about reprocessing Siberian milk not into butter but into cheese, whose main consumers in the West were workers, and for which there was therefore an unlimited market.

In summarizing what has just been said about the absorptive capacity of the market, we could slightly modify our propositions and point out that the organization of marketing on co-operative principles will be easier, the greater the stability of the prices of the product which is made subject to co-operative control.

Apart from the wide capacity of the market and the positive

technical attributes of the commodity in question, the successful co-operative organization of marketing depends to a great extent on the conditions of the commercial organization of the market itself; and on the kind of trade routes along which the commodity travels on its way from the producer to the ultimate consumer.

The production of agricultural goods is usually scattered over an enormous number of small households. An individual household produces goods and disposes of them on the market in small quantities only, and in the case of such products as eggs, poultry, hides, etc., in only a few units or dozens at a time. So far as the pre-war period was concerned, we may note five main stages through which agricultural goods passed on their way through the market.

1. The commodity which was originally scattered among individual producers was collected by a number of buyers and cattle-dealers and became concentrated in their hands.
2. The commodity collected by the buyers underwent a crude grading and was transported from its assembly points to the local centres of wholesale trade.
3. At the wholesale centres, the commodity was further graded and sorted for dispatch to more remote destinations.
4. The commodity was sent on to consumers' wholesale centres.
5. From the consumers' wholesale centres, the commodity reached the consumer via the trade distribution network (local shopkeepers and other traders).

This was the general pattern: it varied greatly for each particular commodity and acquired its own peculiarities.

Thus if, for example, we take a product such as hay, the organization of its market should be regarded as extremely simplified. In most cases, the commodity passed directly from the producer to the consumer; and the middlemen supplying hay to the urban markets were few in number, if they existed at all.

The meat trade – for example, in the meat market in Moscow in earlier times – presented a totally different picture. Livestock which had been fattened in landowners' estates or peasant households was bought up on the spot by cattle-dealers who then transported it to one of the cattle markets in Moscow. At the market, the cattle changed hands among the large-scale traders – the so-called brokers [*komissionery*], who exercised an almost total control over the Moscow market. They re-sold the cattle to the so-called 'bull-slaughterers' [*bykoboitsy*] who slaughtered the cattle and, having separated the remains into carcases, hides and tripe, sold the hides to leather-dressers, the tripe to factories making gelatine and other by-products, and the meat to large and small firms of butchers and to canneries.

It should, however, be remembered, on the one hand, that commercial capital completes not one but several trade turnovers every year; and, on the other hand, that wholesale traders very frequently lend capital on credit both to buyers and to traders. For these reasons, distributors are able to move large consignments of goods through the market by using capital which is worth less than the consignments themselves. This led to an enormous saving of capital; but it also led to a considerable complexity in credit relationships; and those who lent money acquired an extensive power and influence over the entire market and its life.

We have seen how in the past, both in the grain trade and in certain other kinds of trade, large wholesale firms gave credit to – and literally enslaved – local traders who would collect a commodity, giving credit in their turn to the producer and to those commercial consumers (factories, shops, and so on) who then conveyed the product to the ultimate consumer who also relied on credit.

Thanks to the intervention of banking capital in the trade turnover, the dependence of local traders on wholesale dealers at the centre has been considerably reduced; and the market for many goods has ceased to be a virtual monopoly. Thus, for example, the grain market had previously been totally dominated by very large firms. But later, immediately before the war, thanks to the availability of bank credit and elevator technology, the small-scale entrepreneur got the opportunity to operate, and to do so autonomously.

The nature of the market's financial organization, as well as other peculiarities of the market structure, is of enormous importance for the organizers of co-operative marketing.

It is obvious that when setting out to organize the co-operative marketing of a given product, we have to be clear about how the product makes its way through the market in general; and we must ascertain beforehand the obstacles which have to be surmounted as well as the market conditions which can promote the success of co-operation. When making this study, we believe that it is essential to focus our main attention on the following questions:

1. It should be clarified, first of all, to what extent monopoly conditions exist in the part of the market which it is intended to organize on co-operative principles; and how sharp is the competition in this market between buyers and sellers.

A high degree of monopoly is an obstacle to the organization of the market on co-operative principles. An increase in the number of buyers and sellers will tend to favour co-operative marketing. In other words, the organization of the market on co-operative principles will be easier, the less it is dominated by monopoly and the greater the competition on that market between buyers.

In many markets, for example, those for poultry or pigs, it is the largest type of wholesale trade which is most dominated by monopoly. However, this cannot be described as a general rule, since in the case of many commodities, the local markets where the commodities are assembled are more dominated by monopoly than are the large wholesale markets. As a rule, however, the world market is the one which is least dominated by monopolies.

From this it follows that the organization of the market on co-operative principles will be more successful if introduced in precisely that part of the market where the conditions of competition are most favourable to co-operation.

2. In relation to the organization of marketing on co-operative principles, credit conditions are no less important than those of competition.

Credit facilities vary enormously for different commodities and in different parts of one and the same market. During the first stages of the collection of a commodity, all transactions are usually conducted in cash; and we sometimes even have cases where credit is given by the buyer to the seller. In wholesale markets there is the phenomenon of commodity credit, which is given by the seller to the buyer, as well as bank credit advanced on the security of goods. During the stages when the commodity is being distributed, it is extremely common for wholesalers to give credit to retail traders. Therefore, both at the stage when a commodity is being collected and at the stage when it is being distributed, there is the phenomenon of large-scale commercial capital giving credit to its clients; while this commercial capital, in its turn, draws on bank credit.

However, the availability of credit facilities varies for different commodities. Whereas in the cases of cotton and grain credit is widely available in all its forms, nevertheless in the cases of poultry, cattle and flax, credit operations have been little developed.

In the case of co-operative marketing, which is entering the market as a wholesaler, although not as yet fully established in the financial sense, it is obvious that work will be easier, the more widespread is the practice of cash settlements on the market.

3. The question of credit is closely bound up with the question as to how much capital is available to commercial enterprises engaged in marketing. There are some commodities (grain, hides, and so on), whose world wholesale markets involve enterprises which possess

millions of roubles of capital. It is obvious that an infant co-operative movement can compete with such commercial giants only when all other circumstances are exceptionally favourable to it. And in the remaining cases it is obvious that the organization of the market on co-operative principles is easier, the less the amounts of capital being used on the market by private wholesale traders.

4. A not unimportant factor for co-operatives is the rapidity with which a commodity travels on its way through the market. Given the relative monetary weakness of co-operatives, any delays can be exceedingly damaging; and, therefore, commodities whose turnover is slow will be easy to organize on co-operative principles only if credit on the security of goods is readily available.

5. Finally, the techniques required for trading should not be overlooked.

A co-operative official who is not very experienced and who cannot, as a matter of principle, engage in trickery may not be successful for quite a time. Therefore, we would be absolutely right to assume that the parts of the market organization which are most easily organized on co-operative principles are those where trading techniques are the most elementary.

Such are the main conditions of organization which determine the success of marketing on co-operative principles. Besides these, there are certain aspects of production which are of great importance for co-operative marketing. Thus, for example, in the case of most commodities the level of demand is by no means the same throughout the year: there are seasons of heightened demand and seasons where demand declines. It is of the greatest importance that the main bulk of the product should be sent from the households on to the market via the co-operative at a time which coincides with heightened demand. If this dovetailing is not achieved, then co-operatives will face obstacles which will be difficult to circumvent. Thus, for example, co-operatives that sent milk to Moscow – where demand reaches its peak during the winter months – experienced many failures because their members kept to the practice of spring calving – so that their milk supplies were largest in the summer, that is when demand was at its lowest.

It is therefore clear that the organization of the market on co-operative principles will be easier, the greater the dovetailing between the time when a co-operative receives a product from its members and the time when demand for the product is at its height. If this condition is unfulfilled at the time when a co-operative is launched, then the success of co-operative work will depend on the flexibility response of the households who belong to the co-operative and on the speed with which they are able to adapt to market requirements.

In general, the flexibility of response of the households who belong to the co-operative is a powerful factor in the work of organizing the market on co-operative principles. This is because the combination, within a single institution, of wholesale trade and the guidance of production provides an opportunity of following market requirements and of acting with maximum speed to improve the quality of a commodity, and arrange for its primary processing or grading. This can sometimes fetch very high prices on the market and can win an excellent reputation for the co-operative 'trademark' as well as a regular clientele.

But as well as possessing flexibility, households must also possess an adequate degree of stability. They must at all times be able to guarantee to supply their co-operative with goods of a specific and stable quality – upon which the co-operative managers can base their trading calculations and commitments.

In short, we would hardly be mistaken if we were to say that the organization of marketing on co-operative principles will be easier, the greater the flexibility of response, as well as the stability, of the households which have joined in the co-operative.

Finally, it is scarcely necessary to remind our co-operative members of the most important condition of all for their success, without which everything else may prove to be ineffectual. This condition is known to every co-operative member. Its essence is the co-operative's awareness, loyalty and staying power, which enable infant co-operative undertakings to survive the difficult time when they take their first hesitant steps and make their first unavoidable mistakes.

However, while recognizing the exceptionally great, and irreplaceable, power of co-operative awareness and loyalty, we must emphasize that this social phenomenon evolves gradually on the basis of the principle of the direct responsibility of the organs of management to their members, a principle pursued over many years in the internal development of co-operatives; and on the basis of the trust which consequently develops.

But even in these conditions, co-operative organizers should never expose 'co-operative' loyalty to unnecessary strain. Still less can one appeal to such a loyalty when work is just beginning, when marketing co-operatives are at the early stage of development and when the masses still regard them with great scepticism.

6

The Basic Principles of the Co-operative Organization of Commodity Circulation

The organizational problem which faces the peasants when they undertake purchasing and marketing on co-operative principles is a simple one and it has been clearly identified by our earlier argument.

Under a co-operative system, the function of the commodity circulation in the national economy will remain exactly the same as it was before. All that is necessary is that this function should now be carried out by means of a different economic organization. A private commercial apparatus, which successfully ensures the circulation of commodities on the basis of the interests of the commercial capital invested in that circulation, has to give way to a co-operative apparatus. This organization performs the same task; but it is guided, not by its own self-contained interests, but by the interests of those peasant households which created it and which seek, with its help, to resolve the problem of organizing their commodity economy.

When setting up a new economic organization in the place of an old one, the peasant co-operators naturally had to rely on the age-old experience and practice which had evolved through the methods of the old commercial apparatus.

We know that commercial capital was able – as the result of many centuries of working experience – to devise for every kind of market and for every economic situation precisely the kind of commercial organization which could most successfully serve the requisite commercial purposes with the greatest economy of resources.

It is, of course, no accident that in one market a commodity passed on its journey through only one pair of hands, while in another market it passed through five pairs of hands. Nor is it an accident that in one market we find a large number of small traders with little capital, while in another market we find a limited number of traders each of whom, however, has considerable sums of money. Nor again is it an accident that in one case capital circulated rapidly, while in another case goods remained in the same hands for more than a year.

Each of these special circumstances was due to the nature of the commodity and of the market; and the commercial apparatus adapted its structure even to the smallest of these special circumstances.

A co-operative organizer who hopes to replace the existing commercial apparatus has to ascertain what function is performed by each component of this apparatus; and he has to decide what kind of co-operative organ will undertake the task of performing this function. The local cattle-dealer will be replaced by the local co-operative. The local wholesale trader will be replaced by the local territorial association, and the export office by the association at a higher level, as can be seen from Figure 8. As a result of this, the previously existing commercial apparatus will be replaced by a co-operative apparatus, with the same degree of labour specialization and with the same co-ordination of activities.

It can easily be understood that co-operative organizers should not

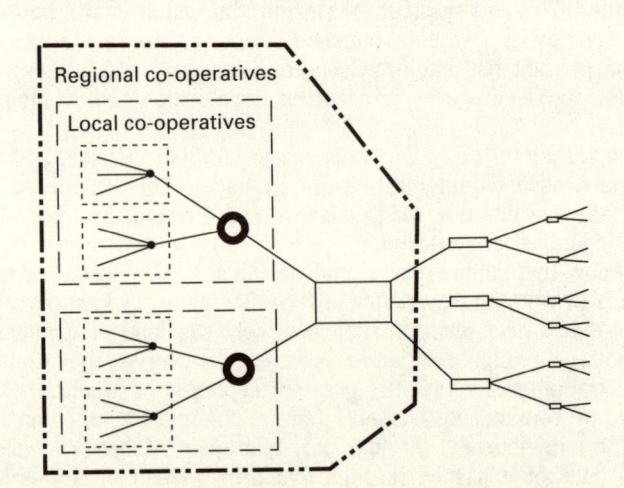

Figure 8: Outline of the organization of the market apparatus on co-operative principles

slavishly imitate the commercial apparatus – which was not itself an ossified or immutable phenomenon, but which evolved with the changing economic situation. But different issues will require still greater changes. The local cattle-dealer and the local credit association are so unalike in their nature that the whole design is bound to be affected when an organizer turns from the former to the latter.

There is much that a co-operative can do more easily and more cheaply than a private trader. Conversely, a buyer is able to do many things which are entirely beyond the capacity of a co-operative organization. Therefore a co-operative organization, when it begins to take hold of the general idea of a trading apparatus – an idea that has developed over the ages – can bring about substantial changes in its forms.

Organizational forms are also bound to be affected by the fundamental difference between the nature of co-operative buying and selling and the nature of private trade. The difference, as we have already seen, is that co-operatives never conduct pure commercial operations, that is, they never make a purchase with a view to re-selling at a higher price. A purchase with a view to re-sale, which constitutes the essence of a commercial operation, is motivated and dictated by the difference between the purchasing and selling prices. For commercial capital, the absolute level of prices is of no very great interest.

The opposite is true of co-operative work which is undertaken either on its own or else for the purposes of joint purchasing or joint marketing. A co-operative apparatus engaged in this work does not, and should not, have any interests apart from those of the peasant households which created it. Therefore, in relation to the co-operative commodity circulation, the absolute price level assumes an exceptional and unique importance. Co-operative marketing must be organized so as to ensure that the peasant receives the highest possible prices for the products of his labour; and co-operative purchasing must provide the peasant with good quality products at the lowest possible prices.

In theory, it is not difficult to solve these organizational problems by using the techniques of the commercial apparatus; and it is logically possible to devise a number of methods and co-operative systems which are capable of organizing joint purchases and joint marketing. Practice shows, however, that co-operative forms evolve in an historical and not in a logical fashion. It often happens that the most subtly devised and profoundly thought out organizational forms will collapse as soon as they make contact with real life.

Any co-operative system for joint marketing or purchasing will show its virtues when resting on the loyal support of those who have united within the co-operative; and when resting on a conscious co-operative discipline which does not permit peasant co-operators to buy or sell outside the co-operative apparatus. But we have to recognize that the Russian peasant masses are far from possessing such a clear awareness. The co-operative milieu is not very cultured; it is far from being aware of its own interests and is often dependent on the local traders. Nor is it receptive to co-operative propaganda. It buys through the co-operative, or entrusts the co-operative with the sale of the products of its labour only when, by so doing, it sees an immediate material advantage in comparison with a sale at a bazaar or a purchase in a shop.

Co-operatives, therefore, have to compete with commercial capital not only in the struggle for wholesale markets abroad, where they contend with the most powerful firms in the world; they also have to compete at the local level for the attention of their own members. Co-operatives can win this latter struggle only if – consistently and from the very start – they offer the peasantry conditions for buying and selling which are more advantageous than those offered by the private trader. They can achieve this only by perfecting the technical and organizational standards of the co-operative apparatus as an *enterprise*.

In theory, all forms of co-operative purchasing and marketing are economically sound and free from the risk of losses. But in practice this proposition holds good only if the co-operative apparatus which is in the process of being established is able, as a commercial enterprise, to stand on at least an equal footing with the very largest commercial enterprises in the same field; and if it can overcome all the difficulties in its work which we have just examined.

It is, at the same time, absolutely essential that a co-operative enterprise, which is capable of acquiring all the competitive power of commercial capitalism, should be developed so as to take full account of the special features of a co-operative association of hundreds of thousands of small peasant households; and that it should carefully follow the basic co-operative principle of the direct responsibility of the organs of management to these peasant masses.

The choice of co-operative forms is, of course, arrived at not as the result of logical analysis but as the result of life itself. And the choice is usually made at the cost of the ruin of numerous co-operative entities which were, in this respect, flawed.

A graphic and very instructive case in point is the history of the most classical type of purchasing co-operation, namely consumer co-operation. The epoch when it first originated is strewn with the corpses of undertakings which, although extremely attractive, were nevertheless weak from an entrepreneurial point of view. The co-operative idea was sound in itself. But it could not become a reality until – through the example of the society in Rochdale – it was able to find specific and viable methods for embodying the idea in an economic enterprise.

Attempts to organize joint purchases during this period usually originated in the form of a mass movement in the pursuit of an idea; and from the organizational point of view, these attempts were based on a very primitive and even naively conceived idea of joint purchasing. Their resources came from credit advanced by customers; the consignments of the commodity which had been acquired were distributed at cost price; and out of philanthropy, credit was widely made available, throughout the entire operation, to the poorest members.

The first stages of such an undertaking usually gave an impression of stupendous success: prices fell and all the consumers remained satisfied. However, a reaction very quickly set in. Local shopkeepers, irritated by the fall in prices, embarked on a life and death struggle, using all their capital for the purpose of trading below cost, thus undercutting the co-operative undertaking and depriving it of its customers.

These efforts nearly always succeeded and friction then arose among co-operative members. Members who had borowed money were unable to repay it. Unexpected losses, damage to goods and enormous overhead expenses came to light; and, at the first jolt on the market, the enterprise collapsed like a house of cards.

It was in this way that the initial English and other similar undertakings met their end. It was obvious that any direct application of the principle of joint purchasing would be totally unable to withstand the onslaught of life and that it was, from the entrepreneurial point of view, utterly unsuitable.

An obvious need was felt to find organizational forms of work of a kind which would, without abandoning the movement's ideological goals, nevertheless produce solid and organizationally stable forms of co-operative enterprise. In 1845 these forms were discovered in the enterprise of the 'Just Pioneers'. One has only to recall the basic principles of the Rochdale weavers in order to grasp their very profound importance and vitality, specifically from the entrepreneurial point of view.

Thus the basic Rochdale principle lays it down that commodities bought by a co-operative at low prices on the wholesale market must be distributed between the co-operative's members not at their commercial cost but at the customary average prices on the retail market. This principle is put forward to counter-balance the usual practice of rudimentary joint purchases, where a commodity is bought up wholesale out of money which has been collected and is then distributed to the buyers at the wholesale price with a surcharge for overhead expenses.

The Rochdale principle, which might seem to contradict the principles of joint purchasing, results in an enormous strengthening of the co-operative as an enterprise in the following ways:

1. The co-operative is thereby enabled, over the course of the year, to consolidate its usually meagre turnover capital.
2. It is enabled, out of its profits, to cover any expenditure on organization over and above what it had budgeted for; to cover any fortuitous financial or material losses; and to sell some goods below their purchase price.
3. It can make purchases on credit, although this is usually less profitable.

In other words, the possibility of earning a significant, even though short-term, profit within a co-operative apparatus makes it more stable and more flexible from the entrepreneurial point of view.

The second Rochdale principle envisages that all profits earned from this additional charge over and above the cost of the products should be repaid at the end of the year to those who invested in it, that is, in proportion to their annual contribution to the buying of the goods.

This principle restores the economic profitability of joint purchases, and indeed strengthens profitability by acting as a source of savings. Thanks to the delay in refunding these additional payments for a whole year, the ten- and twenty-kopeck pieces put aside every day are recorded day-by-day and combined into a single sum which ultimately grows into an amount of real importance in a worker's budget.

The annual payment of such a relatively large sum may attract the attention of the peasants and serve as the best kind of co-operative propaganda.

The third rule, which proclaims that it is not permissible for a co-operative to adulterate or give short weight for a product, stemmed from the principle of joint purchasing. Co-operatives, since they are interested in the absolute price of the product obtained and not in the difference between the purchase and resale prices, have to avoid adulteration and dilution, if only because the work involved in adulteration adds to the original cost of the product which is obtained and consumed by the peasant.

There is, however, no doubt that the practical adoption of this principle weakened the position of co-operatives as enterprises and deprived them of a number of economic opportunities.

The first of these lost opportunities had been that of trading on credit. The entrepreneurial calculation which leads a private shopkeeper to bestow extensive 'favours' on a buyer by allowing him to have goods on credit, rests on two foundations:

1. By selling on credit, private traders expand their turnovers, thereby reducing overhead costs and increasing the volume of profits; and
2. When a buyer becomes indebted to a private trader, he makes himself dependent on the trader and can easily become the victim of exploitation. Income derived from adulteration, from giving blatantly short weights, from charging higher prices and from the selling of shop-soiled goods will more than offset the unavoidable proportion of bad debts which always occur when trade is conducted on credit.

But the adoption of this Rochdale principle means that in the case of co-operatives, the latter consideration totally disappears; and a purchasers' co-operative, by forgoing price surcharges, deprives itself of the only available means of covering the inevitable losses which result from granting credit to impecunious buyers. Therefore, from an entrepreneurial point of view, it is precisely the co-operatives which consider the granting of credit to be impermissible. These were the reasons which led co-operators to adopt the fourth principle of the 'Just Pioneers', even though it meant an inevitable curtailment of their operations. This fourth principle laid it down that in a co-operative shop, transactions can be conducted only in cash.

This principle was also upheld on the grounds that if credit were widely made available to members, the co-operative's meagre capital would become inextricably tied up in loans; and the co-operative would have extremely little power on the wholesale purchasing market. Not having any liquid resources, it missed profitable trading opportunities and was obliged to rely on burdensome forms of credit provided by commercial capital.

Therefore, in those cases where social conditions prompted co-operative organizers to think about giving credit assistance to their poorest members, they tackled the problem by creating special credit funds; and they based their credit facilities entirely on these funds, so as to ensure that losses in no way affected the entrepreneurial foundations of joint purchasing.

We can therefore see that three out of the four basic principles of the Rochdale weavers are derived, not from the foundations of co-operative ideology,[1] but from the entrepreneurial considerations of the co-operative organizer.

These same entrepreneurial considerations also gave rise to the fifth basic principle of purchasing co-operatives, which is expressed in its financial structure. So far, this has not crystallized sufficiently clearly to be defined in a succinct slogan.

In cases of rudimentary joint purchasing, the peasant household does not set aside any separate or special resources. Each peasant household advances the money which it would have to pay if it made the purchase itself. The money collected is pooled and used to buy commodities which are then handed over to the households which contributed to the purchase. However, this method of financing operates only when joint purchases are made at irregular intervals. In so far as joint purchasing becomes a regular operation, the regular collection of money and the organization of the purchasing itself will develop on a scale where it becomes profitable for peasant households to delegate joint purchasing operations to a special enterprise and provide this enterprise with its own resources.

There is no doubt that purchasing co-operatives will cope more efficiently with their tasks, if they function as an enterprise. But at the same time, the existence of these co-operatives in this form will require the peasants to make advance contributions to the turnover capital of this new co-operative enterprise over and above the enterprise's expenses which the peasants paid before. This expenditure could be very substantial.

If, let us suppose, a peasant makes purchases worth 300 roubles a year, and if the capital in the co-operative circulates ten times per year, then the peasant must make an advance contribution to the co-operative's capital of 30 roubles, because only in this case will the co-operative be able to provide him with its services.

This is a very substantial sum; and the peasant will be able to pay it in addition to his ordinary expenses, only when he feels sure that the saving from the joint purchase is substantially greater than the amount of his share in the co-operative.

Therefore, the principle has been urged that the capital of a purchasing society must be financed out of the savings made by joint purchasing and must not be financed out of the pocket of the peasant. In other words, when a purchasing co-operative begins its work it must – through the collection of direct share contributions from its members – build up capital of a small amount which is not burdensome for those who pay. During its first years of trading, it must rely on credit from commercial firms. Then, having achieved a substantial saving as a result of the difference between the cost of procuring goods and the average market prices at which it is required to sell them in accordance with the first Rochdale principle, the co-operative must, in accordance with the second Rochdale principle, repay to the contributors only a part of this saving. The remainder of the proceeds must be used by the co-operative to build up its capital, which thus accumulates as its work proceeds.

Such are the basic principles of the Rochdale Consumer Co-operatives. According to these principles, the process of joint purchasing is developed as follows:

1. A special enterprise is created, on the basis of capital separate from the peasant household; and it organizes its own apparatus for purchasing and distribution.
2. The co-operative apparatus thus created uses its share capital and the credits which it has obtained to buy commodities of which it becomes the owner as a juridical person.
3. The commodities are distributed between the members; but this is done through a change of their legal ownership, by purchase and sale at average market prices and for cash.
4. The profit earned by the co-operative enterprise as a result of the difference between procurement value and the price when the distribution is made is partly invested in the enterprise's capital and partly repaid to the co-operative's members in proportion to what they have contributed.
5. The management of the enterprise is undertaken by elected collective organs and in accordance with the rules drawn up at meetings of the members, whose rights are determined by their membership and not by the amount of their shares in the co-operative's capital.

Such is the co-operative organization which evolved historically for the purpose of solving the problem of joint purchasing, and which, from this point of view, is very successfully tackling the organization of the peasant household's money economy.

In view of the abundance of literature on the organization of consumer societies, we have confined ourselves to the general considerations just set out; and we shall entirely omit any detailed description of the co-operative purchasing apparatus.

Co-operation for the purchase of the means of production for peasant households – such as seeds, machinery, fertilizers, and so on – is organizationally separate from consumer co-operation but forms part of the general agricultural co-operation which we are examining. Its organizational foundations are identical to those explained above; and if they differ at all, they differ mainly in the sense that purchases are made at more irregular intervals. For this reason, advance contributions are more often collected from those taking part in a purchase. This similarity of organizational principles makes it unnecessary for us to provide any special description of purchasing co-operatives.

Far more basic and important is the distinction that separates purchasing co-operation from consumer co-operation, with regard to the purpose for which commodities are handed over to the peasant household.

Purchasing co-operatives, which supply the peasant with the means of production, impose upon the co-operative apparatus not only the task of conducting commercial operations, but the further task of making important judgements of an agronomical nature as to the qualities of the goods that they provide. A purchasing association, which supplies rural areas with seed, has to guarantee that they are capable of germinating and are economically suitable; and it naturally acquires the function of offering guidance as to what varieties of seed the peasants should sow in their fields.

In these conditions, part of the organization of production is shifted by the individual peasant household on to a collective enterprise. In order to perform this role, the organization of purchasing will acquire certain special characteristics. The link between purchasing co-operatives and peasant households will become closer than in the case of consumer co-operatives.

The same is true with regard to the choice of machinery and even the planning of new types of machinery. All this will give purchasing co-operatives the role, not so much of centres for supplying households with the mechanical means of production, but rather, the role of centres for collective thinking, which organize the means of production within the peasant household, both through the organization of joint purchasing and also through other methods of influencing the peasant household. It is, indeed, not without reason that purchasing co-operatives become surrounded by experimental stations for seed and machinery, by laboratories and by other institutions engaged in working out methods for agricultural production.

These are the reasons that lead us for the most part to classify purchasing co-operatives as a form of producer co-operation. Some of the leaders of purchasing co-operatives have recently been inclined to regard them, not as co-operatives engaged in providing supplies, but as co-operatives which organize the means of production of peasant households – thus still further emphasizing their importance for production.

Even more closely linked with production is co-operative marketing which organizes another aspect of the peasant family farm's money economy.

The Basic Principles of the Co-operative Organization 125

From a logical point of view, the problem of co-operative marketing can be solved as simply as that of joint purchasing. Peasants who want to sell the products of their labour at wholesale prices will assemble small amounts of the commodity which they produce – such as flax, eggs, milk or hemp – into sufficiently large commercial consignments and, by offering them on the wholesale market, they find advantageous buyers.

However, the numerous failures of co-operative undertakings in this field show us beyond doubt that the concrete realization of the simple idea of co-operative marketing proves to be an even more difficult task than that of joint purchasing.

In the first place, the peasant, oppressed by a constant shortage of money, will be very reluctant to wait for the co-operative to collect the necessary consignments, sort them through, sell them and then hand over the proceeds of sale to the peasant two or three months later. The need for money is generally so acute that peasant householders prefer to sell the flax straight away, even if at lower prices.

For this reason, it is hardly ever possible to conduct a simple joint marketing operation in a pure form. It is essential to pay something to the peasants long before the products collected from them have been sold, in order to satisfy their desperate need for money.

Marketing co-operatives cannot, however, follow the path of consumer co-operatives, by creating a co-operative enterprise which buys the peasants' products out of the capital built up through share contributions. This is impossible if only because, in view of the seasonal nature of sales, this capital has to be equal to the annual turnover or, what comes to the same thing, it must be equal to the peasants' annual receipts for the product which is jointly marketed.

It is clear that the payment of such substantial sums of money is beyond what the peasant household can afford; and co-operative marketing is usually organized in the form of a rudimentary operation for joint selling combined with a parallel operation for granting credit to peasant households on the security of the commodity which they have provided for the joint sale.

Under these conditions, the peasant's commodity, throughout the whole time when it is handled by the co-operative apparatus, remains the property of the peasant, which he has handed over to the co-operative on commission for the purpose of being sold.

To put it more simply, the peasant who has brought his products for marketing is given a certain sum of money, equal to part of the estimated value of the products which he has brought. But this money cannot be deemed to be a partial payment for the product. It is merely a loan secured by the value of the product to be marketed; and is distinct and separate from the marketing operation.

The loans provided on the security of the commodity in question are usually equal to one half, two-thirds or three-quarters of its estimated value, but with the aim of ensuring that if prices fall, the fall in the value of the product does not make the security worth less than the loan. This sum of money is usually quite sufficient to sustain the household until the product is realized. Unfortunately, however, our co-operative members are still far from possessing the necessary co-operative awareness and staying power; and since they very often believe that a bird in the hand is worth two in the bush, they will take their product to the bazaar if the cattle-dealers in the bazaar pay more than the amount of the pre-payment, i.e. the initial loan offered by the co-operative.

In order to combat this kind of phenomenon, many co-operatives follow the wrong path: instead of developing their members' co-operative awareness, they begin to increase the size of loans in relation to the value of the secured goods. They offer loans equal not to three-quarters of the goods' estimated value, but equal to 80 per cent, 90 per cent, 95 per cent or even 100 per cent of their estimated value.

It is true that in theory, even with loans of this size, the lending is still not the same as a final purchase of the goods because, in the event of the goods being sold at a higher price, the peasant receives an additional payment, while in the event of a sale at a price below the value of the loan, the difference must be recovered from the peasant who took out the loan. This is provided by the co-operative's usual rules. But it is quite obvious that these theoretical considerations are only valid in relation to the supplementary payment; because it is hardly possible to count on recovering any sum of money already paid to the peasants without thereby undermining the co-operative undertaking itself.

Co-operative practitioners are themselves well aware of this; and in order to neutralize the effect of these virtually 'final' purchases they often resort to devices which have very little in common with the co-operative spirit. Thus in one popular book on co-operative marketing we read that:

> Our flax cultivators, who do not understand the rules of co-operative marketing and who are accustomed to selling goods at the bazaar for a fixed price, will demand that co-operatives should also pay for the product in full. In such cases, some co-operatives, if their members prove stubborn, will adopt the following method: they pay their members for the flax in full, but at the same time they either undervalue the flax or assess it at half a grade below what it merits.

High payments when goods are accepted remain a scourge for co-operative marketing. It has to be remembered that co-operative marketing means precisely the joint marketing of a product by its producers; in no sense does it amount to trade in the sense of 'buying for the purpose of subsequent resale'. This constitutes the whole strength of co-operation, because it operates not on the speculative difference between buying and selling prices but on the full value of the product, which means payment for the labour put into it; and therefore it cannot entail a net loss, no matter how much prices may fall. The transition to operations based on a final price exposes co-operatives to the risk of losses and undermines one of their most stable foundations.

We think it necessary to note, however, that a large number of practitioners of the co-operative movement point out the desirability of gradually creating adequate capital reserves and of building up marketing co-operation on the same model as that of consumer co-operation – that is, by paying the peasant a final price for the goods which the co-operative acquires from him, and distributing the co-operative's profits at the end of the year in proportion to the value of the goods handed over for marketing. This new principle, together with its theoretical justification, has recently been advocated with great fervour by A. Chizhikov. He points to the greater simplicity and convenience for the peasants of having their goods bought at market prices and being given receipts entitling them to possible further payments, in preference to the ritual of secured loans and commissions which are often pointless and irritating for the peasantry. We shall not dispute the substance of these arguments. They are undoubtedly valid for those co-operatives that have become established and have accumulated a reserve capital out of their profits. The whole question is whether the time is yet ripe for Soviet co-operatives to adopt such a system.

For there can be no doubt that the transition to operations of this kind will only be possible when confidence exists that the difference between buying and selling prices will at all times be at least enough to cover the cost of maintaining the co-operative. This – and indeed the very existence of marketing co-operatives in general – will be possible only if the co-operative apparatus has achieved a high technical and organizational standard; and this latter condition can be fulfilled only if co-operatives enter the market from the very start as powerful large-scale centralized organizations, whose management possesses virtually dictatorial powers.

It may be asked why we do not insist on such drastic organizational methods in the case of the other types of co-operative – such as consumer and credit co-operatives.

In order to clarify this question, we need, in both cases, to grasp the economic nature of the process which is being organized on co-operative principles; and we need to realize that in the case of marketing, the effects of centralization, managerial decision-making power and scale are much more significant when expressed in quantitative terms, than they are in the case of purchasing.

We have to remember that the victory of co-operatives is achieved, not by logical argument, but through an intense economic struggle and through the conquest of the market as a result of the efforts of co-operative organizations. That being so, we can point to two crucial differences between, for example, consumer and marketing co-operatives.

First, so far as consumer co-operatives are concerned, the first stages of their work do not in any way involve technical reprocessing, grading, packing or anything else to do with the preparation of the goods for the market, since the co-operative acquires the goods on the wholesale market in a technically ready-made form. However, in the case of co-operatives for the marketing of raw materials, the primary co-operative unit, which replaces the cattle-dealer and buyer, must not only assemble the goods but must carry out the work of technically grading and processing them. It is precisely this aspect of work which is especially difficult for a co-operative in its early stages of existence. And it is precisely here that the guiding hand of the centre – equipped with a high degree of technical experience and able to make the fullest use of science as well as the whole range of technical information – is alone capable of coping with the task in hand. For consumer co-operatives, however, the task of reprocessing usually arises during the later stages of work, rather than at the beginning.

Secondly, what is perhaps even more important is the fact that in the case of a transaction by a consumer co-operative, the decisive stage comes in the small, local retail markets in competition with the small retail trader whose turnover is comparatively limited and whose economic power is slight.

Therefore, a consumer society, even during its initial development, will be large enough to be competitive and can compete successfully against small local traders. And an association at the district level may, even in its initial phase, immediately become a large and powerful firm, invulnerable to competition from the small village trader.

Marketing co-operatives are in a different position. The decisive stage for them when they enter into a transaction occurs on the wholesale market, when they come up against economic giants who possess knowledge and economic experience gained over many years and who also dispose of large amounts of capital and even larger credit facilities. There are a whole number of goods – hides, for example – where the only consignments of any importance for the market are valued in hundreds of thousands of roubles.

Therefore, except in the case of a small number of markets which are specialized by their nature – such as the markets for milk, vegetables, flowers, poultry, and so on – a small co-operative which enters the wholesale market with its consignments of goods will be a supplicant rather than a seller.

To sum up what has just been said, we can outline the following organizational scheme for marketing co-operatives, based on the usual principles of intermediary activity and the granting of secured loans:

1. A special enterprise is set up, with share capital contributed by the peasants, which is enough to maintain the apparatus of the marketing co-operative.
2. The organization set up in this way acquires from the peasants the products of their labour for sale on commission. These goods, after a preliminary valuation, are graded, made uniform and assembled into commercial consignments.
3. At the same time as a commodity is accepted on commission, it is pledged by way of security to the usual co-operative credit apparatus which advances a loan equal to part of the estimated value of the commodity.
4. After the commodity has been sold on the market, the proceeds of its sale are used to meet the loan advanced by the credit apparatus together with the percentage commission to cover the costs of the co-operative marketing apparatus. The remainder of the proceeds of sale, after the deduction of a certain sum which is used to build up the co-operative's own capital, is handed over to the peasants who brought in their goods for sale on commission.
5. A co-operative marketing enterprise is set up in the form of a large centralized organization, managed by elected collective organs in accordance with the rules laid down at the meeting of the members, whose rights are determined by the fact of their membership – and not by the proportion of shares which they have contributed to the capital of the co-operative marketing enterprise.

Such is the co-operative organization which has evolved under the pressure of everyday life and which has proved to be stable.

One has only to compare the five points of its organizational scheme with the corresponding points which we formulated for consumer co-operatives in order to understand the organizational differences between them. Such a comparison will demonstrate that a consumer co-operative as an enterprise is a good deal more detached from the economic activity of its members than is a marketing co-operative. It must be noted, however, that even when marketing co-operatives are strengthened from an entrepreneurial point of view and change over from the system of intermediary activities and the granting of secured loans to a system closer to that of the Rochdale principles – that is, to the Chizhikov system – their links with peasant output cannot disappear. This is because the links between peasant output and the organization of the peasant economy are extremely close, owing to the fact that the interests of joint marketing require that a commodity should be of high quality and should correspond to the demands of the market.

A co-operative organization will naturally try to exert all the influence it can to re-organize the economy so as to meet these demands. Flax co-operatives will make every effort to bring Russian fibres up to the requirements of English flax mills. Co-operatives for the marketing of eggs and poultry will try to organize their output in accordance with the demands of the corresponding markets. And since co-operatives are able to rely not only on advice and practical example, but also on the powerful incentive of offering higher payment for goods of a desirable grade, one need have no doubt that their influence will be effective.

Co-operative principles have been penetrating production as a result of the organization of the market on co-operative principles, in the same way that commercial capitalism in its time paved the way for industrial capitalism.

NOTE

1. We hasten to point out that all the principles of the Rochdale pioneers themselves had an ideological motivation which is now forgotten and which does not form part of the ideology of present-day co-operative activists.
   [Editor's note: For a recent history of the Rochdale Pioneers, see Co-operative Union, *Rochdale Pioneers Memorial Museum, The House of Co-operation* (The Co-operative Press 1990).]

7

The Organization of Co-operative Marketing and Reprocessing Enterprises

In order to assemble wholesale consignments of a commodity, it is obviously necessary to rely on large areas which contain scattered peasant settlements. And it is necessary, first of all, to collect the commodities by transporting them across these areas to local centres.

This trading region must not be too small, since in that case its trade turnovers will be insufficient to pay for the equipment and maintenance of the co-ordinating centre. On the other hand, the trading region must not be too large, since in that case, the transport costs and the physical labour involved in moving the commodities will be disproportionately great. In general, the apparatus that assembles the goods can cope with the transport required for the preliminary collection only if the peasant sellers themselves are made responsible for bringing the product from their households to the preliminary assembly point. For this reason, the area of the original market, that is the size of the trading region which surrounds this co-ordinating centre, has been historically determined by the peasants' transport facilities – by the possibility of making a return journey by horse and cart within a single day, and so forth.

In an historical perspective, these trading areas evolve gradually: they constitute, as it were, the molecules of the national economy and serve as the basis for all kinds of administrative and territorial divisions (such as the *volost'* [sub-district], and so on). In recent years, these areas of commercial gravitation have been the subject of very detailed statistical investigation. By showing on one map the places where the peasants settled and the centres to which they delivered their products, these investigations have given us a number of vivid geographical pictures of the territorial organization of this initial collection of goods.

In order to get a complete picture of the organizational problem that faces us in regard to the initial collection of agricultural raw material, we have to focus attention not only on the spatial organization of the market but on its organization over time. Different products of agricultural labour are produced in peasant households at different times; and because of varying marketing conditions as well as varying degrees of the need for money, these products are delivered by farmers to the market with varying degrees of rapidity.

The existing data on this question indicate that peak periods of delivery to the market are August and September in the case of grain; December and January in the case of flax; and May and June in the case of deliveries of fresh milk in the area around Moscow. Meat deliveries to the market vary according to variations in the fodder base: in the case of grasslands, deliveries are largest in the summer; in the case of regions that grow sugar beet, deliveries are largest in the winter. It has to be noted that these typical sales curves do not remain constant from year to year: they change under the pressure of the economic situation.

When small consignments of a product are made up into large commercial consignments, they become intermingled and are then reprocessed and regraded; that is to say, they cease to have any identifiable physical connection with their owner and they become *depersonalized*. For this reason the owner-producer no longer has any right to the specific article, but is entitled only to its value. Therefore the value, or at least the quality, of the product has to be precisely ascertained when the product is accepted; and this must be recorded in the documents handed to the owner.

The fair and accurate valuation of the commodity represents a tremendous moral asset for co-operatives. It is also a highly effective way of influencing the way that peasant households are organized. Recognition of the superior quality of a product, and the readiness to pay a higher price for this quality, will make the farmer aware of the requirements of technical progress. It is no accident that when co-operatives in the Vologda province measured the fat content in milk, this led to the deliberate rejection of certain herds of cows owned by peasants and to the selection of cows according to productivity. This holds out the promise of a major transformation in dairy farming in the north over the next decade; and it also prepares the way for a wideranging development of specialized associations of producers.

In order to provide maximum encouragement for technical progress in peasant households, the co-operative apparatus can even go slightly beyond its reliance on market valuations: it can award bonuses to the products and grades which it needs by making additional payments over and above the market price. These additional payments can do much to encourage improvements already planned. They must not be unduly high since the peasant household is usually very sensitive to price changes. Therefore, from the financial point of view, the bonuses can be paid for without any difficulty – even, perhaps, by offering reduced valuations for those types of goods which are either due to be discarded, because of the planned agricultural programme, or which are very inconvenient for the co-operative because of the difficulty of selling them on the market owing to lack of demand.

A consignment of goods, once it has been accepted from the peasant and given a valuation, is then sent off to be marketed. It would be quite conceivable in theory – and it would indeed be the best thing from the economic point of view – for the peasant to refrain from demanding any immediate payment when he hands over the produce of his household to the co-operative; and for him to wait until the end of the operation to receive the money earned by the co-operative for the sale of his products.

However, because of the population's level of understanding and constant lack of financial resources, the peasants will very seldom offer their goods for sale on such terms; and when they hand over their products they usually want to be paid at least a part of the goods' value. Therefore, side by side with the co-operative marketing system – and the words 'side by side' should be particularly stressed – a system of co-operative credit is organized on the security of the commodity which is sent for marketing.

The peasant who brings his products for marketing is given a certain sum of money, equal to part of the products' estimated value. But, once again, this sum of money cannot under any circumstances be regarded as a part-payment for the product. It can be regarded as nothing more than a loan, which is separate and distinct from the marketing operation and is secured by the value of the product sent for marketing.

In order to explain the exact difference between the earlier system of co-operative marketing – of intermediary activities based on the granting of secured loans – and the new system which has now been adopted by the majority of butter-producing co-operatives, we would offer the following point-by-point comparison of the organizational principles in each case (see table 20).

There is no doubt that the second system considerably simplifies the co-operative's office work and gives it a free hand with its commodity operations. At the same time, it does involve the possible

Table 20

System based on intermediary operations and secured loans	System based on final purchase of goods
(1) A special enterprise is set up, provided by its peasant membership with a small amount of capital necessary and sufficient to maintain the co-operative marketing apparatus.	(1) A special enterprise is set up, based on special share capital and borrowed capital, sufficient not only to maintain the co-operative marketing apparatus but also to produce a trade turnover.
(2) The apparatus thus set up receives the products of the peasants' labour for the purpose of sale on commission. These products are given a preliminary valuation and are then depersonalized, graded and used to make up commercial consignments.	(2) The co-operative apparatus thus set up on the basis of the share capital and credit facilities which it obtains, acquires the products of the peasants' labour through the legal process of purchase and sale at market prices and for cash; and the co-operative apparatus becomes the legal owner of the products.
(3) At the same time as the commodity is accepted for sale on commission, it is pledged to the usual credit co-operative, which advances a loan equal to part of the estimated value of the pledged commodity.	(3) No credit operation takes place. The valuation under the preceding paragraph is made in full on the basis of a final price. However, a record is kept of each member's delivery of products and of its amount.
(4) After the commodity has been sold on the market, the proceeds of sale are used to pay off the loan advanced by the credit apparatus and the percentage commission which covers the costs of the marketing co-operative apparatus. The remainder of the proceeds of sale are all transferred to the peasant who handed over the commodity for sale on commission.	(4) After the sale of the commodity which has been assembled, the profit obtained by the co-operative apparatus from the difference between the procurement and selling prices is partly used to increase the enterprise's capital and partly repaid to the members who handed over their goods, in proportion to what each of them handed over.

risk of losses, which are in theory ruled out in the case of co-operatives engaged in intermediary operations based on secured loans. It also involves the risk that members will be less interested in the affairs of their co-operative as an enterprise.

However, supporters of the second system maintain that when loans equal to 100 per cent of a commodity's estimated value are advanced, we are in fact dealing with exactly the same kind of operation for final purchase, which is merely complicated by the remnants of the already outdated middleman's commission. And indeed, if large amounts of money are advanced on security, there is in fact little difference between the two systems, since, if the goods are resold at a loss, it is hardly possible to recover the money advanced without destroying the authority and reputation of the co-operative.

After a product has been sorted through, it must, in order to become a commodity, undergo a certain amount of processing of the kind required by the market; and it must be packaged. Our co-operators often devote too little attention to packaging. However, it is an important factor in success both in preserving the commodity and in making a favourable impression on the buyer. For these reasons, packaging techniques must be given the highest possible priority.

The problem of determining the selling price is in effect the problem of determining the highest price at which the co-operative commodity which has been assembled can be sold on the market.

In order to ascertain the highest possible prices, co-operative leaders have to analyse the current state of the market as well as the possible development of the factors which determine the formation of prices. The basis for calculating prices is the valuation given to goods when they are accepted, combined with the overhead trading expenses borne by the co-operative apparatus. These two amounts added together represent the limit below which prices must not be allowed to fall. However, the level of this particular 'cost' should be regarded as only the lower limit of prices; and the whole art of those who handle the transaction is to try to raise the bargain price as much as possible above this level.

Only a central organization that is in a position to follow the state of the markets and the changes occurring within it is capable of finding the price level which it is seeking.

Records of prices in the consumer market in present and past years, records of stocks in the hands of consumers, producers and commercial middlemen, the state of harvests and the level of well-being among the peasants who possess the product – these, and scores of other factors have to be taken into account when ascertaining price levels.

When the average price level, which is usually calculated for the average brand of a given commodity, has been sought after and ascertained, we then have to tackle the second part of the problem and break down the general price of a commodity into the prices for its particular brands.

In the case of a whole number of agricultural products – butter, eggs, meat, poultry and even bread – it is vitally important to be equipped for the conservation of the products. There is hardly any need in this book to mention the importance of elevators and refrigerators for co-operative marketing. This question has been so often discussed in co-operative literature that one may assume that the importance of elevator and refrigeration equipment is fully appreciated by co-operatives.

Less attention has been given to the problem of the conservation of goods in transit even though this is crucially important for a whole number of products. It is enough, for example, to say that the export of butter abroad during the summer is possible only with the careful provision of refrigeration equipment and packaging on the railway journey.

Even less attention has been paid to our usual warehouses; although the successful provision of enclosed storage space can significantly reduce overhead expenses. It is not difficult to build the usual type of warehouse with a large amount of space. It is much more difficult to organize matters so as to make do with the minimum amount of enclosed space whilst using the available space to the fullest possible extent.

It has been the usual practice of co-operatives to develop three types of warehouse.

1. Warehouses for the reception of products at assembly-points situated near bazaars or local co-operatives. The commodities do not remain for long in these warehouses, but are continually being moved to:
2. Warehouses managed by the co-operative associations and situated near railway stations, from which the commercial consignments already made up are sent to:
3. Central warehouses or warehouses situated in ports.

The accumulation of large quantities of goods entailed a considerable risk of loss in the event of fire. For this reason many organizations preferred to use different warehouses even within one and the same town, thus spreading the risk and protecting themselves against losses through fire.

So long as marketing co-operatives remain weak – selling their goods without any charge for storage or dispatch and sending only negligible consignments – questions of transport or transport packaging are of no great urgency. But as soon as marketing co-operatives grow to levels of world-wide importance and begin to dispatch large quantities of their product, on a scale which is significant even in the context of the national economy, the problem of transport becomes increasingly urgent. It becomes essential to prepare co-ordinated plans for transport and storage; to work out the shortest distances; to fight for low fares and low freight charges after carefully studying this question; and to look for ways of minimizing transport costs.

In order to underpin the power of co-operatives in relation to transport, it was necessary to set up a special transport centre, a kind of commissariat of communications. This centre, which handles an exceptionally large volume of freight, had to bring co-operative transport up to the required standard, both in regard to domestic communications and also in regard to maritime export routes.

The problem of supplying a commodity to customers is the problem of the commercial policy of the co-operatives in question. If co-operatives wish to take the maximum advantage of the prevailing market situation, then they have to pursue an active, rather than passive, commercial policy.

The first requirement for success from this point of view is a system of complete and rapid information as to the state of the market. The export organization must always keep its finger on the pulse of the world market and be sensitive to the slightest changes. It must rapidly get its bearings in a changing situation and, while acting in accordance with the situation at the given moment, must not lose sight of what the future may hold in store; and must be able to anticipate market trends.

Such information requires the intensive use of specialized resources. It can only be obtained by a central organization, which is able to set up a research department – a kind of observatory to monitor the state of the market.

Using the data from such an observatory, a co-operative marketing centre must identify its most regular and reliable customers, and must also, as far as possible, identify the customers who are closest to the ultimate consumer. It must try to turn them into its regular clientele.

The organization of permanent commercial ties of this kind is an immensely important task; and for this purpose, co-operatives can and must forgo all kinds of dubious transactions which sometimes offer large profits, but profits in the short term only.

It must never be forgotten that co-operatives are designed to last, not for one year but for centuries at the very least; and thought has to be given to the co-operative's well-being over a period of many years and not just to its advantage in any one year. This is why a regular clientele is one of the surest foundations of success in the entire work of co-operative marketing. By gradually studying the requirements of regular customers, by learning their wishes, by winning their confidence in the quality of co-operative produce, it is possible to achieve mutual harmonization of interests and benefits. For the producer this means getting a high reward for his labour. For the consumer, it means a high quality product, produced even in its initial stages so as to conform with the consumer's requirements.

At the present time, a system of regular clienteles is organized by agricultural co-operative centres and associations, which enter into general contracts with trusts, syndicates and other purchasing organizations. Provided that they are entered into on a basis of equality, we regard contracts of this kind as a grain gain.

Co-ordination between consumers and producers has been achieved to a high degree by the egg co-operative system in Denmark, where every consumer is able, from the number displayed on the egg, to find the actual producers and convey praise or criticism to them.

Permanent commercial links of this kind may be confirmed by special preliminary agreements or by contracts between the parties; but must not, of course, restrict the right of the organized producer to defend his interests.

We pointed out at the beginning of this book how the right to enter into a transaction gradually passed from the local association to the central organization. Another question of very great importance is the question of exactly who, within the central organization itself, should have the right to enter into final transactions.

This brings us to a fundamental question relating to the organizational-administrative development of co-operation.

The nature of co-operative institutions entails a collegial form of management. The collegial bodies which run the affairs of co-operative institutions are the general meeting, the council and the board of management; and there is no doubt that only their proper functioning can guarantee that the will of the co-operative truly reflects the will of the majority of its members.

A proper implementation of the principle of collegiality will lend stability to what the co-operative does, it will produce cohesion among the co-operative's membership and it will make the nature of the work clear to all who take part in the organization.

However, a co-operative is not only a self-sufficient democratically structured organization. It is also an economic enterprise operating in the conditions of the capitalist world. The advantages of collegiality from the point of view of the co-operative's internal structure are often outweighed by major shortcomings from the point of view of the business requirements of organizing an enterprise.

Every businessman in the capitalist world who manages an enterprise on his own, will run the enterprise by closely monitoring the pulse of economic life, by making rapid and flexible organizational adjustments in response to changes in the economic situation; and to a great extent grasping what needs to be done by relying on hunch.

A co-operative organizer is in a different position. Because the co-operative is managed on a collegial basis, he must not only satisfy himself that a particular step is sound and advantageous, but must unfailingly satisfy his partners on the collegial bodies. Persuasion, above all else, takes time. Often, by the time that the collegial bodies have been won over by the argument, the favourable economic situation has vanished irretrievably and the step which had been contemplated ceases to have any purpose.

It need hardly be pointed out how greatly this weakens the position of co-operatives in their struggle against private entrepreneurs who proceed mainly through action and above all through action without discussions or consultations.

Thus, while the nature of co-operation calls for the principle of collegiality in management, the interests of commercial success call for the principle of individual decision-making.

Life has to reconcile these two principles. In fact, it is the principle of individual decision-making which predominated in the largest and most successful co-operative undertakings. But it was maintained not on a formal basis, but on the basis of authority and personal trust.

It is, of course, difficult to offer any prescriptions for such a complex and controversial problem. But it seems to us that collegial bodies should confine themselves to giving general instructions as to the aims of work; and that they should leave the task of implementation to the individual decisions of those whom the collegial bodies have chosen as their agents. The collegial principle should apply to the election of officers and the approval of their plan of work; but the work itself should be left to individuals. In cases where such an agent finds that it is beyond his capacity to carry out a particular decision, he should inform the collegial body.

Such is the basic organizational problem. The success of co-operative work will itself depend on the successful solution of this problem.

After repaying all borrowed money to credit institutions – this repayment being deemed to have satisfied all loans taken out by the peasants on the security of goods – the central association has to pay the surplus funds to the producers, after deducting its own percentage commission and after covering its own trade expenses as well as the commissions and expenses of local associations and co-operatives.

Financial settlements with the peasants may be made on the basis of the final purchase of the products which they deliver to the co-operative. This system does not exclude the possibility of supplementary payments. But in this case, the supplementary payments will be made to the peasants by way of a distribution of co-operative income, and not by way of a handing over of the proceeds of sale after the payment of commission and other deductions.

During good years, the amount of these supplementary payments may be as high as 20–5 per cent of the original valuation of the commodity; and the possibility of these supplementary payments undoubtedly strengthens co-operative discipline. Nevertheless, co-operative organizers who hand out millions of roubles for these supplementary payments, cannot help thinking that at least some of these millions of roubles, instead of being scattered among individual peasant households, ought to be accumulated within co-operative organizations in the form of indivisible capital.

For marketing co-operatives the most important capital of this kind is 'guarantee capital' which is built up to cover losses which may result from a disastrous fall in prices; and, in some cases, for the purpose of selling their commodities below their 'cost' in new markets.

Capital of this kind may serve a number of other purposes – such as insurance, the organization of production, educational activities, and so forth. In any case, all capital of this kind enhances the strength of the co-operative association and thus enhances the strength of the peasant farmers on the world market.

The formation of these kinds of capital is far more important from the national point of view than might at first appear. The co-operative apparatus, by eliminating the private commercial middlemen, thereby considerably reduces the capitalist profit which is formed in the country's national economy; and it achieves a democratization in the distribution of the national income. The national income ceases to be concentrated in the hands of a small number of people; and is distributed amongst tens of millions of small producers. But despite the positive significance of this process from the social point of view, one is bound to recognize that this income, if scattered among millions of owners who have no connection with each other, will be consumed to a considerably greater extent than in a capitalist society; and therefore the country's capacity for capital formation will be lessened by this democratization. Therefore, for the purpose of compensating for this phenomenon, and of increasing the capacity for capital formation, the formation of the peasantry's social capital mentioned above is a matter of special importance.

Financial settlements with the peasants mark the end of all direct co-operative marketing operations. In addition to these, however, co-operative institutions inevitably have to take a number of measures for the purpose of changing the organizational principles of the peasant economy itself, which constitutes the foundation of this co-operative structure.

This activity is, it is true, only at the stage of being planned; but it is objectively essential and therefore inevitable in a wide-ranging form.

Marketing co-operatives are interested in obtaining from the peasants only those raw materials which are greatly in demand and which can easily be sold on the market. Therefore co-operatives naturally make every effort to persuade the peasants to adopt methods of cultivation which actually produce the kind of raw material required by the market. And since, in most cases, the labour applied to such raw materials is more generously remunerated by the market, such persuasion – if backed up by higher valuations when products are accepted – cannot fail to influence the peasants and lead to the progressive reform of the peasant household.

In this area of its activity also, the peasant co-operative system changes – imperceptibly but very profoundly – from an organization of middlemen into an organizer of production. And it would probably be no great exaggeration to say that the influence of marketing co-operatives on the organization of the peasant economy is no less important, and may perhaps be more important, than are our publicly sponsored projects relating to agronomy or the work of our special agricultural societies. The level of remuneration for the peasant's labour when he sells his products is very often the most powerful factor which influences his economic effort. And in many cases where the most impassioned sermon about agriculture does nothing to win over the peasant mind, the valuation of products accepted for co-operative marketing proves a more effective method, which does not even attempt to get the peasant's conscious agreement to the reform being carried through. An even more powerful factor will, of course, be the combined influence on agricultural reform of all the resources which the peasant co-operative system has at its disposal.

Such are the basic organizational elements in the process of co-operative marketing which we have been analysing. The work of turning them into a system of concrete activities, of co-ordinating these activities over time and space and choosing executive staff – all these are tasks which depend on the personal skill of the organizers and leaders of the co-operative movement.

THE CO-OPERATIVE MARKETING OF AGRICULTURAL PRODUCTS COMBINED WITH REPROCESSING

Primary reprocessing is, generally speaking, an integral part of the peasant household's activity. For example, the threshing of crops and the separation of the wheat from the chaff effectively involve an element of primary reprocessing. In the broad sense of the word, primary reprocessing includes all mechanical processes which alter the form either of the produce which has been harvested, or of the produce of cattle-rearing. Examples of such reprocessing are: the skimming of cream from milk, the manufacture of butter, cottage cheese or soured cream [*smetana*], or the curing of tobacco. The considerable importance of such reprocessing can be demonstrated from the example of flax cultivation. It can be seen from Table 21 that out of 82.2 days spent on cultivating a hectare of flax, 44.9 days are spent on threshing, braking and scutching, that is, in the primary reprocessing of the product.

Table 21 *The organization of the work of flax cultivation (number of working days per hectare)*

	Days	%
Two ploughings and two harrowings	8.2	10.1
Sowing	2.4	2.9
Picking the flax	14.1	17.5
Spreading and removing the flax from the fields	12.6	15.2
Drying and threshing	10.0	12.1
Braking	14.5	17.6
Scutching	20.4	24.8
Total	82.2	100.0

Thus 54.2 per cent of the work of flax cultivation is taken up by mechanical reprocessing and only 45.8 per cent is taken up by working the soil, care of the land and harvesting.

If we take not only labour-intensive flax cultivation but also field cultivation, we find that in an average household in Volokolamsk the various forms of primary processing of flax, grain and clover account for 39.3 per cent of all work.

The mechanical methods of primary reprocessing can – unlike

biological processes – easily be detached from the organizational plan of the individual household, and can with considerable advantage be organized as large-scale production. Second only to marketing, procurement and credit, primary reprocessing is perhaps the most promising sector of an agricultural enterprise, for the purpose of demonstrating the advantage of a large-scale form of production.

It is, therefore, quite natural that when the peasants organized marketing on co-operative principles they also sought to extend the co-operative system to the primary reprocessing of the products which they were marketing.

At the same time, however, it should be noted that the practitioners of co-operatives did not confine themselves to organizing those types of primary reprocessing which already existed in individual peasant households (butter manufacture, scutching of flax, and so on). They also began to introduce types of reprocessing which were beyond the capacity of individual peasant households and which had never previously been undertaken although they were necessitated by the requirements of marketing operations. These included:

1. Reprocessing which made a commodity easier to transport and thus expanded the geographical areas in which co-operative marketing operations could be carried on. This included the reprocessing of potatoes into starch and treacle, the drying of vegetables, and so on.
2. Reprocessing for the purpose of preserving highly perishable products. For example, the canning of fruit and vegetables, the salting or refrigeration of meat, the preparation of dried eggs, cheese-making, etc.
3. Reprocessing which enabled the owner of a raw material to earn the high income derived from the reprocessing industry – for instance, the milling of grain, the reprocessing of oil, the preparation of tobacco, and so on.

Undertakings like these are so obviously beneficial they have long since become commonplace in the co-operative movement. In the two following chapters we shall look at the way these ideas have been embodied in practice.

However, before we turn to these particular relationships, we have to resolve the basic question as to what kinds of activity are suitable for peasant co-operatives engaged in reprocessing. Specifically, the most important problem when organizing co-operative reprocessing is to decide up to what stage this reprocessing should be taken. Thus, for example, flax co-operatives can confine themselves to setting up scutching centres, they can build mechanical equipment for flax-combing, they can set up their own flax mills or even build weaving factories and supply the market with the produce of their fields in the form of finished products. So far as potato production is concerned, co-operators can confine themselves to setting up potato-grinding factories manufacturing raw starch. But they can take their reprocessing even further by manufacturing processed starch, treacle or even treacle jam, sweet-meats and spice cakes. Ought we to try to take all these stages of reprocessing into our own hands? Or should we confine ourselves to a narrower range of co-operative production and limit ourselves to the manufacture of semi-finished products?

A capitalist organizer has an incentive to set up a projected enterprise if, and only if, he calculates that this may bring him a high revenue. But for us, the co-operators, the question is a good deal more complex. We must not forget that our basic task as producers, as well as in all other spheres of co-operative work, is not to obtain the highest possible profit from enterprises which we have set up on the basis of hired labour, but to use these enterprises so as to increase the remuneration for our own labour.

It is unacceptable for co-operative reprocessing to be turned into the capitalist exploitation of hired labour. The peasant who has become a co-operator must not turn into a capitalist. The overwhelming bulk of the labour expended on the products which he sells on the market must be the labour of the peasant co-operators themselves.

This rule is based not only on ideological but also on purely practical economic considerations. As we demonstrated in earlier chapters, the main competitive strength of marketing co-operatives stems from the fact that by working, not with purchased goods but with their own goods, co-operatives can withstand any fall in prices without disturbing their own apparatus, by shifting losses on to the peasant household itself which possesses tremendous stability and flexibility.

This advantage makes co-operatives invulnerable in the struggle with commercial capital; and it is an advantage which co-operatives also possess when dealing with raw material produced within the peasant household. But the advantage is noticeably diminished when the raw material is subjected by the co-operative to significant reprocessing, organized on capitalist principles; and when the value of other people's labour and the financial costs begin to make up a significant part of the value of the product sold. In that case all these advantages will eventually disappear and the co-operative enterprise will find itself in the same economic conditions as those that apply to commercial and industrial capital.

A co-operative, since it is always less enterprising and rejects in principle many forms of commercial work, will find it extremely difficult to compete with commercial and industrial capital in the organization of capitalist production. For these reasons, any excessive enthusiasm for reprocessing will undermine our positions.

Such are the considerations which oblige peasant co-operatives to confine themselves only to those stages of reprocessing where the value of the raw material constitutes the overwhelming part of the value of the finished product, thus maintaining all the advantages of a co-operative enterprise over a capitalist enterprise.

This golden rule should not be forgotten by today's peasant farmers. It is precisely the mineral substances – so important for the soil – which largely end up as technical waste material when reprocessing takes place. Therefore, a peasant who wants to maintain the fertility of his land must not lose these minerals; and for this reason he must organize all the various kinds of reprocessing in such a way that such by-products remain in the hands of peasant co-operatives.

These are the purely agricultural considerations which make it necessary, other things being equal, to leave peasant co-operatives in control of cream production, the production of sugar beet, and flour-grinding. A number of prominent agronomists headed by A. Minin consider this to be imperative.

Having thus set out the various considerations which are usually taken into account when organizing one or other kind of co-operative production, we should note that the reprocessing of a particular product may quite often be appropriately undertaken, not only by agricultural co-operatives, but by other kinds of co-operative.

This inevitably raises the question of possible competition between different co-operatives. Those who support the unity of the co-operative movement would like co-operative production to provide a basis for the fusion of co-operatives of all types – whether run by workers, peasants or townspeople. They would also like to organize industrial enterprises as joint co-operative ventures. However, such a fusion does not seem to us to be feasible. Peasant and urban co-operatives may be able to avoid mutual enmity. They may, furthermore, be able to enter into commercial transactions with each other. They may even achieve unity at congresses or in common organizations, based on a common idea or formed for financial or insurance purposes. But the producers' interests and the commercial interests, which it is sought to protect, are so utterly opposed that they cannot be unified within a single organization, because the will of such an organization will inevitably be undermined by internal contradictions between the opposing interests which have been combined into a single whole.

Therefore, we think it is better not to talk about organizing production in factories and mills through the joint efforts of different types of co-operatives, but to talk instead about demarcating different sectors of production between these co-operatives. The criteria for such a demarcation exist, and earlier pages of this book have shown in great detail why agricultural co-operatives can control those types of production where agricultural raw material has a preponderant importance and where waste material can be used for agricultural production.

We would willingly leave all remaining types of production to consumer co-operatives, since in their case, the co-operative principle is maintained not through production but through ensuring stability of demand.

From a logical point of view these propositions, like many others, are beyond dispute. However, economic entities come into being not in a logical, but in an historical fashion. Therefore, it is in the sphere of production above all that we shall see the most marked competition between different types of co-operative system. Mills, treacle factories and tobacco factories, and so on, operated respectively by consumer and peasant co-operatives, will exist at one and the same time; and the question of what belongs to whom will ultimately be determined by the course of economic history. No planned regulation will be able to eliminate this struggle between different systems.

Our enterprises for the primary reprocessing of agricultural products at present constitute the area of co-operative work which most urgently needs attention. It is precisely these enterprises which provide us with the broadest opportunities and most fascinating prospects; and it is also these enterprises that give rise to the greatest doubts and apprehensions. The success of butter manufacture co-operatives and the numerous failures of the flax-processing factories require particular study.

8
Machinery Users' Associations

The butter-producing partnership, the consumers' shop, the marketing association and the credit association organize on co-operative principles most of those elements in the peasant household's organizational plan which link the family farm with the outside world.

It is precisely this enlargement of the scale of economic turnover that has brought the most visible and the most profitable results for the peasant household while leaving its individuality virtually intact. It was here that the process began of organizing the peasant household on co-operative principles; and it is here that co-operation has been most extensively developed. However, this still leaves the possibility that there may be certain other processes, inherently connected with the internal economic activities of the family farm, which can also be organized on co-operative principles – and which can bring an appreciable benefit to the household, while likewise doing nothing to destroy the individuality of the peasant household's remaining sectors.

A thoughtful observer who examines the organizational plan of the peasant family farm can see that it involves a good many technical processes where an enlargement of the scale of operations can bring a considerable profit. Furthermore, the practical experience of co-operatives points to a good many cases where co-operative principles have penetrated into the very core of peasant production and into areas of the household's economic activity which have no connection with the market.

The most characteristic example of this kind of co-operation is the so-called machinery users' association. This enables a small self-employed peasant farm [*trudovoye krest'yanskoye khozyaistvo*][1] to make use of complex machinery which can only be profitable when substantial areas of land are being worked or when large quantities of a product are being reprocessed. In order to get the clearest possible notion of their economic nature and of their importance for the organization of small peasant farms, we have to analyse the general problem of the mechanization of agriculture.

From the economic point of view, the question of the mechanization of agriculture fundamentally hinges on calculating the cost of mechanized work – both in absolute figures and in comparison with the cost of the same work done by hand.

Among all the literature in this field, the simplest, clearest and most distinctive treatment of the question which concerns us here is to be found in a small book written by the French agronomist, F. Beçu. In this book, which consists of a study of labour organization in the French department of the Pas-de-Calais,[2] Beçu produces the following general formula for calculating the cost of mechanized work.

Let us, so Beçu says, use the symbol A to denote the amount of annual expenditure incurred on a machine, irrespective of whether the machine is used or whether it remains idle, i.e. the expenditure on amortization, the interest on the capital invested in the machine and the insurance premiums.

Let us then use the symbol B to denote the costs of employing mechanized labour for each day that the machine is used, i.e. the wages of the workers operating each machine, the cost of traction and the costs of oiling and repairs.

The cost of mechanized work per day will then be expressed by the symbols:

$$x = \frac{A}{n} + B$$

where n equals the number of days per year when the machine was in use.

It is clear from this formula that the cost of one day of mechanized work will be lower, the higher the variable n, in other words the more fully the machine is used.

In order to ascertain the cost of work done by a machine per hectare of land, it is necessary to divide the resulting figure for the cost of work per day into the number of hectares per day which the machine services.

If the daily productivity of the machine is equal to k hectares, then the cost of mechanized work per hectare will be equal to:

$$y = \frac{\frac{A}{n} + B}{k}$$

A machine often performs only part of the work, leaving the remainder to be done by hand (for example, the operation of reaping machines requires manual work by the binders). Therefore, if we use the symbol C to denote the cost of this residual manual work per hectare of land, we can abbreviate the formula above so as to obtain a final expression of the cost per hectare of work done by a machine as follows:

$$z - \frac{A}{n.k} + \frac{B}{k} + C$$

If, on the basis of Beçu's formula, we now use the symbol R to denote the cost of manual work per hectare, we can confirm that the replacement of manual work by mechanized work is profitable if:

$$\left(\frac{A}{n.k} + \frac{B}{k} + C \right) < R$$

Let us now try to examine the conditions which are necessary to produce this disparity:

On the left-hand side of this formula are the quantities denoted by A, B, C and K, which depend on the cost and quality of the machine and on the level of wages. These amounts are relatively stable and constant. The amount subject to the most variation is the one denoted by n, i.e. the number of days during which the machine can be used during the year. This amount depends on the area of land at the disposal of the household; and the amount denoted by $n.k$ in our formula is a direct expression of this all-important land area.

If the area of land used ($n.k$) decreases, the costs of amortization and the interest on the capital (A) will relate to a smaller number of hectares; and as a result of this the cost of mechanized operations will substantially increase and will exceed the usual cost of manual work. In order to determine the limit of the area of territory on which the use of the machine is possible, i.e. the area in which the cost of work done by a machine is equal to that of manual work, we have, in our formula, to use the symbol x to denote the amount $n.k$ which we are trying to discover; and we have to write the following equation:

$$R = \left(\frac{A}{x} + \frac{B}{k} + C \right)$$

From which we derive:

$$x = \frac{A}{R - \left(\dfrac{B}{k} + C\right)}$$

It is obvious that if x is greater than this amount, then the cost of mechanized operations will be less than R (the cost of operations by hand), whereas if x is less than this amount, then it will be greater than R.

We shall now explain these theoretical calculations by reference to the specific example of the operation of a mowing-machine. Let us suppose that its productivity covers a land area of 3.5 hectares per day, that it costs 200 roubles and that the basic components of our formula work out as follows:

Amount A

4% on capital	8	roubles
Amortization (over 10 years)	20	"
Total of amount A =	28	roubles

Amount B

Wages of worker	1	rouble
Cost of traction (with two horses)	1	rouble 50 k.
Oiling and repairs	1	rouble
Total of amount B =	3	roubles 50 k.

C is equal to zero since the mowing is entirely mechanized.

The mowing by hand of 1 hectare per day requires three workers. If they are each paid 1 rouble per day, then the cost of the manual operation is 3 roubles.

Let us suppose that a household possesses 70 hectares of meadow. In that case, according to our formula, the cost of the mechanized reaping of 1 hectare is as follows:

$$z = \frac{28}{70} + \frac{3.50}{3.5} = 1 \text{ rouble } 40 \text{ kopecks}$$

Thus, given 70 hectares of meadow, mechanical reaping is more than twice as profitable as reaping done by hand.

Let us now use our formula in order to ascertain the minimum land area on which it is economically possible to use a mowing machine. We find that:

$$x = \frac{28}{3.00 \; \frac{3.50}{3.5}} = 14 \text{ hectares}$$

Thus mechanized mowing is profitable only in households which possess at least 14 hectares of meadow.

Let us take, for example, a household which has only 7 hectares of meadow. In this case, the cost of mechanized mowing of 1 hectare will work out as follows:

$$z = \frac{28}{7} + \frac{3.50}{3.5} = 5 \text{ roubles}$$

that is, it will cost 2 roubles more than mowing by hand.

In all the preceding tabulations and calculations, we have assumed that mechanized agricultural work will be of the same quality as work done by hand. But this is not so in reality. We know that when sowing is carried out with a seed drill, we achieve not only a saving of labour but an economy in sowing material and we save 6–8 poods of seed per hectare. Furthermore, sowing in drills increases the crop yield. We also know that when a threshing-machine is used, the speeding-up of the operation reduces the amount of grain which is devoured by mice – although straw, on the other hand, suffers more damage from a threshing-machine than it suffers from a flail. We also know that to use modern harrows not only speeds up the work but increases the harvest yield, and so forth. We obviously need to make allowance in our formula for this effect, translated, of course, into roubles. If we express the improvement (or worsening) of the quality of work as a result of mechanization in terms of N roubles per hectare, we can calculate the cost of mechanized work compared to manual work in the following way:

$$\frac{A}{n.k} + \frac{B}{k} + C - N \gtrless R$$

On this basis, the limit of the area for the profitable use of machine can be expressed in the form:

$$x = \frac{A}{(R+N) - \left(\frac{B}{k} + C\right)}$$

In the case of some machines (such as the seed drill and the wooden plough) the land area calculated in this way will be less than in the case of a calculation which disregards the qualitative factor.

The situation examined above is due to the well-known fact that the use of machinery declines in proportion to the diminution in the area of the farm. Thus, for example, in Germany before the [First World] war the percentage of farms using machinery in relation to the overall total of farms is shown in Table 22.

Table 22 *Influence of the size of farms on the dissemination of agricultural machinery*

Size of farms (ha)	Percentage of farms using machinery
0–0.5	0.9
0.5–2	8.9
2–5	32.4
5–20	72.5
20–100	92.0
100.0 or more	97.5

It must, however, be pointed out that the formula just explained is an immutable law only for enterprises organized on capitalist lines. The ideas that underlie the organization of a self-employed peasant family farm will often result in substantial modifications of this law.

Thus, for example, in the south of the USSR at the present time, the use of reaping-machines and even of binders has become widespread in the peasant economy. Furthermore, these machines are being operated by farms whose land area is so insignificantly small that, according to our formulas, the operation of machines cannot be profitable to these farms. Therefore, in this case, the reasons for their widespread use have to be sought not in their profitability but in the special characteristics of a self-employed peasant farm.

One of the problems of such a farm, which distinguishes it from a farm organized on capitalist lines, is that of spreading its work as evenly as possible over time. Capitalist farms, which obtain their manpower on the free market for labour in accordance with the requirements of their organizational plan, are able to ignore this requirement. For that reason their organization of manpower usually

Figure 9: Distribution of labour at different times on sugar beet farms in Austria

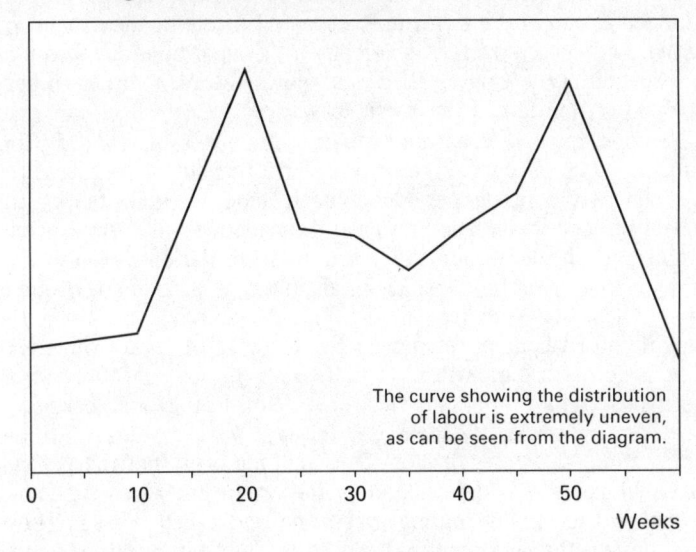

The curve showing the distribution of labour is extremely uneven, as can be seen from the diagram.

Weeks

It can be seen from this figure that the curve showing the distribution of labour is extremely uneven.

Figure 10: Distribution of work in the cultivation of wheat during different months

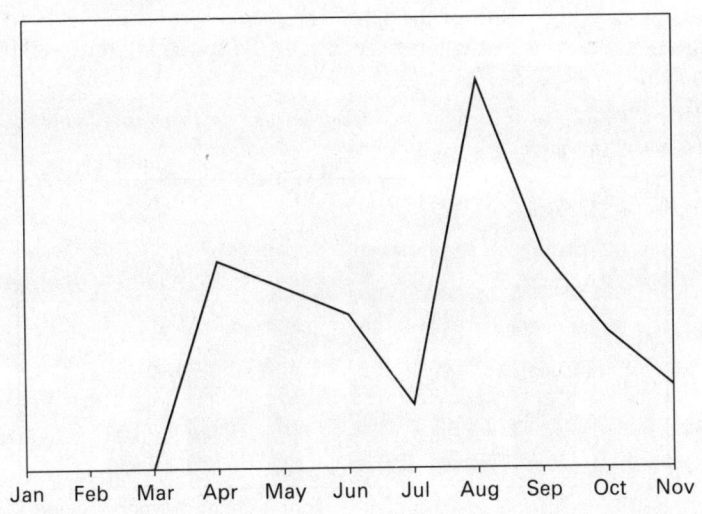

involves an extremely uneven distribution of effort over different periods of the year. Thus, for example, a study carried out by K. Linder of one of the farms in Austria organized for the production of grain, with a considerable development of sugar beet, has given us the following curve (Figure 9) which shows the number of workers employed on the farm from week to week.[3]

But a self-employed peasant farm cannot allow its periods of effort to follow such a curve, since the farm cannot accomplish the necessary work during periods of peak activity, while during the slackest periods it is obliged to leave its manpower idle. As a result, peasant households usually suffer acutely from the uneven organization of labour over time, which is an inherent problem with many crops. Thus, for example, Figure 10 shows how uneven is the distribution of labour in relation to the growing of spring wheat.

The harvest time, which is the peak period of effort, thus determines the area of land which it is feasible to work. If ripened wheat can remain in the field, let us say for 1½ weeks, without deteriorating, then it is obvious that the land area that the peasant household can sow is determined by the area from which the family can gather the harvest during those one and a half weeks. These limitations on the land area that can be exploited have an extremely unfavourable effect during other periods of the year, because during these periods the family is unable, on this limited area of land, to provide employment for all its manpower; and it suffers from a surplus of labour for which there is no work.

When seeking to expand the land area that it can exploit, the peasantry in the south of the USSR sometimes sows its fields with varieties of wheat that can remain in the field for a long time without

Table 23 *Labour needed for production process on 1 hectare of land sown with wheat (in days)*

Ploughing		3.6
Sowing		1.7
Weeding		4.4
Harvesting		4.3
Carting		1.9
Threshing		3.6
Winnowing		1.9
	Total	21.4

deteriorating (for example, *beloturka*). By sowing it instead of other more profitable varieties of wheat, the peasant household consequently reduces its 'net income' per unit of land; but at the same time it gains the opportunity to increase its land-holding worked by family labour and thus to increase its gross income.

The same significance attaches to the use of harvesting machines over small areas of land, where the machines do not repay their cost.

Thus, for example, according to statistical data for the *Zemstvo* in the Starobelsk district of the Kharkov province,[4] the gathering of the harvest from 1 hectare of land requires an outlay of 4.3 working days out of the 21.4 days needed for the production process as a whole.

Let us then suppose that we have a family with two workers and that the period of harvesting can be extended over ten days. In that case, the maximum area from which the family can gather the harvest by its own labour will be:

$$\frac{20}{4.3} = 4.65 \text{ hectares}$$

And since a hectare requires in all 21.4 working days and yields a gross income of 29 roubles 10 kopecks (after deducting the cost of seed) it follows that our family which is engaged in economic management will be able to perform a total of 94.8 days' work (47.4 working days per worker per year) and will be able to increase its means of livelihood by a total amount of 139.3 roubles.

However, by using a reaping-machine, the household can achieve a more than twofold expansion of the area it cultivates; and by sowing, let us say, 10 hectares, it will be in a position to perform about 200 days' work over the year and to earn 291.6 roubles of gross income. If, from this sum, we subtract 30 roubles for amortization and for the repair of the machinery, we arrive at a sum of 261.6 roubles, that is, over 100 roubles more than can be produced by the use of manual labour only. Such a significant increase in the means of livelihood is of immense advantage to a self-employed peasant farm, despite the fact that on a book-keeping calculation, the use of a reaping-machine on 10 hectares of land would undoubtedly make a loss.

This, then, is the significance of machinery in a self-employed peasant family farm in coping with the critical periods when extra effort is required. But the mechanization of labour has an entirely different role during slack periods. Thus, for example, at a conference of agronomists in Perm in 1900, one agronomist, D. Kirsanov, pointed out that:

If during the winter, the labour of a peasant family is profitably employed, an agronomist will do the most good by promoting the distribution of threshing-machines, thus freeing a significant amount of peasant labour for productive activity of other kinds. But if, during the winter, the peasant has nothing to do except thresh his crops, then the dissemination of threshing-machines can scarcely be seen as anything but an unproductive outlay of peasant capital, which is meagre enough as it is.

Kirsanov very aptly points to a case where the aims of a self-employed peasant farm may come into conflict with the mechanization of labour, even though mechanized work may perhaps – from a book-keeping point of view – be extremely profitable.

We shall not give any lengthy description of the managerial bodies of the machinery users' association, nor of the other details of its internal organization, which resemble those of other co-operatives. We shall, however, concentrate on the four most important organizational problems which determine the association's work in relation to the joint utilization of agricultural stock.

The first of these problems concerns the methods of raising capital for the purpose of acquiring the necessary stock. A certain proportion of this capital comes from the share contributions of participants. However, stock can easily be sold off in the event of the association going into liquidation. Therefore, share capital may serve as no more than a guarantee against losses in the event of such a sale of assets; and most of the necessary funds come from long-term loans which are repaid through surcharges over and above the usual charges for the use of the machines. Given the association's solvency, which is in any case guaranteed by the stock itself and by the share contributions, co-operative credit centres can finance them without any misgivings; and can do so even in the absence of any provision in the association's statutes making members liable for the association's affairs.

We know of cases, however, where machinery associations based on loan capital as well as on small share contributions have consolidated their financial position by introducing a limited liability on the part of members over and above the members' share contributions.

After they have, in one way or another, sorted out the financing of the machinery association and obtained the capital needed to buy the stock, the association's organizers then have to resolve the second organizational problem: they have to decide on what principles the association's machinery is to be selected.

Sometimes this problem is settled extremely simply. It is

ascertained in the members' households exactly what kind of work the association can help to mechanize. Then, once the productivity of each kind of machine has been determined, a calculation is made as to how many of these machines are needed in order to carry out the work in question. Thus, for instance, if an association has 40 members and if each member has sown an average of 10 hectares with cereal crops, then the total area served by the machinery association will be 400 hectares. Assuming a harvest yield of 12.8 quintals per hectare, this will provide a total yield of 5,120 quintals.

In order to exploit this area, use may be made of the machines listed in Table 24 which would be beyond the means of small farms acting on their own:

Table 24

Name of machine	Productivity per day	Duration of the season when it is used	Productivity during the season
Seed drill	4 hectares	18 days	72
Disc-harrow	3 "	50 "	150
Reaping-machine	3 "	10 "	30
Threshing-machine (with two horses)	20 quintals	80 "	1,600
Winnowing-machine	40 "	80 "	3,200
Separator	60 "	100 "	6,000

If we now apply the productivity rates of these machines as shown above to the calculation which concerns us here, we get the result shown in Table 25.

Such, in outline, are the calculations. But when they are applied in practice, there are a great many complicating circumstances which have to be taken into account. Thus:

1. Machines, if they are to be used by a large number of small farms, must be mobile, and it takes time to move them. The beginning and ending of the work may very often be such that a machine is used for less than a whole day, which also leads to a waste of time available for the machine's use.
2. If a machine is to be fully employed during the season, people have to queue to use it; and this queue can itself be extended

Table 25 *Number of machines required to service a land area of 400 desyatiny [1,080 acres] sown with grain*

Name of machine	Volume of work	Seasonal norm	Number required
'Randal' harrow	400 hectares	150 hectares	3
Seed drill	400 "	72 "	5
Reaping-machine	400 "	30 "	14
Threshing-machine	5,120 quintals	1,600 quintals	4
Winnowing-machine	5,120 "	3,200 "	2
Separator	5,120 "	6,000 "	1

over the entire season. The longer it proves possible to extend the season for the user, the more fully the machine will be used and the lower will be the cost for each individual farm.

However, if the queue begins too early or ends too late, this will be extremely inconvenient for farmers, because it will upset all their economic calculations. For this reason an association has to cope with a sharply intensified demand at the height of the season; and under the pressure of this demand it has to increase the number of machines for which there is a demand over and above what is required by the norms of profitability. The result is that machinery associations have more machines than are needed, and the machines are never fully utilized.

3. The two circumstances just mentioned both make it necessary to increase the number of machines operated by the association over and above the guidelines set out earlier. There are, however, other considerations which tend to reduce this number. For example, it is quite possible to imagine that many farms do not wish to mechanize all work which is capable of being mechanized and that they will leave many kinds of work, such as harrowing and reaping, to be performed manually or by means of rudimentary implements. Moreover, some of the larger farms will prefer to buy for themselves some of the machines – such as winnowing-machines – which they rent for use on small areas of land. Both these circumstances tend to reduce the number of machines which the association requires.

Thus, the real demand for machinery may not coincide with rudimentary theoretical calculations of such demand. A machinery users' association therefore has to weigh up the considerations just mentioned and it has to decide, gradually and by a process of trial and error, what to include in its equipment.

Usually, however, financial considerations alone will prevent the association from buying its equipment all at once. The machines are acquired gradually, starting with those most in demand; and the association's board of management always has the opportunity to adapt to the requirements of everyday life. It should be recognized, in any case, that for a machinery users' association as an enterprise, a surplus of machines – that is, an extensive underemployment of machines – has been more dangerous than a partial shortage of machines.

So far as the collection of machines is concerned, one is bound to highlight and emphasize the point that an association must, during its first years of existence, be able to provide the most profitable and efficient machines, both in order to build up the association's reputation in the eyes of the peasantry, and also in order to strengthen the association's internal capabilities as an enterprise.

But to follow this rule consistently is not as easy as it might seem. To begin with, machinery users' associations are usually formed at the time when the most profitable machines are already being used, either privately by a large number of farmers or else by groups of two or three homesteads jointly. Even so, these machines are, of course, used only to a very small extent and they do not work out cheaply for their owners. Thus, for example, P. Vikhlyayev, who in 1910 made a study of the flax district of the Moscow province, has written that:

> Provided that threshing-machines are evenly distributed over the land and provided that these machines are all used to the same extent, it requires in all only 8 days to thresh the entire crop in a year with an average harvest. And for particular sub-districts [*volosti*] such as the Kul'pin and Bukholov sub-districts, it takes only 5 or 6 days to put the whole crop through the existing threshing-machines.
>
> The average crop can be winnowed in two days with the winnowing-machines accounted for in the census of 1910.
>
> This results in an actual surplus of grain-harvesting machinery in peasant households. The mainly private ownership of the best equipment tends to produce the same result: it leads to the extreme under-utilization of the machinery. In these conditions, it is not very productive for the population to spend money on acquiring improved equipment.[5]

It is desirable that everything possible should be done to remove such equipment from private use by individuals and transfer it to cooperative use. This will, in the first place, save national economic capital; and it will, secondly, do much to strengthen machinery users' associations.

However, the peasantry's innate individualism and the perceived advantages of private user lead farmers to cling very tenaciously to these machines; and to offer machinery users' associations only the equipment which is either totally beyond the handling capacity of individual farms or is of a kind whose value is unclear and uncertain. A good deal of tact and skilful effort is needed in order to steer an association away from the pursuit of a ruinous course in these matters.

Many machines which are not very well known also turn out to be not very suitable. There is no demand for them on the part of the members; and the machines stand idle and become an unnecessary burden – a kind of junk. Moreover, when machines are of different types, each one of them needs its own particular spare parts and repairing equipment, which immensely complicates repair work. When all the machines are of the same type, they are serviced by the same repair equipment; and where the need arises, or in the case of serious breakages, it may be possible to convert two broken machines into one, which can be operated immediately without waiting for a capital repair. These considerations relating to repair are not only a matter of convenience. They also demonstrate how repairs – and therefore the use of machinery – can be made less costly.

Therefore, in the older machinery users' associations, the variety of types of machine has been gradually reduced; and these associations seek, for working purposes, to collect machines of one type only. This also settles the basic economic task of this kind of cooperative, which is to provide for the joint user of large machines which are beyond the means of small individual farms. The association's economic planning and the whole of its organization have to be adapted to this basic task – which the association tries to fulfil in the most satisfactory way possible.

But a second task – that of carrying out practical tests on new types of machines and encouraging their use by the peasantry – needs to be undertaken separately from the basic tasks. Where special resources are made available for this purpose, they are organized on their own and on different lines, without interfering with the entrepreneurial foundations of the associations' main work. Profits are not uncommonly used to create a special fund for the testing of new machinery as well as a fund for propaganda, and so forth. These have made it possible to conduct testing and propaganda activities successfully and on a very large scale, but in ways which do not interfere with the associations' basic work and which are not motivated by considerations of entrepreneurial gain.

NOTES

1. Editor's note: The expression 'self-employed peasant farm' has been used here to translate the Russian term *trudovoye [krest'yanskoye] khozyaistvo*, which referred to a farm entirely dependent on the labour of members of the peasant family.
2. F. Beçu, *le travail agricole et la condition des ouvriers agricoles dans le Departement du Pas-de-Calais,* Paris, 1909.
3. K. Linder, 'Die zeitliche Verteilung der Handarbeit in der Landwirtschaft', Tiel's *Landwirt Jahrbuch,* Vol. 38.
4. *Materialy dlya otsenki zemel' Khar'kovskoi gubernii* [Data on Land Valuations in the Kharkov Province], Volume III, Kharkov, 1907.
5. P. A. Vikhlyayev, *Vliyanie travoseyaniya na otdel'nye storony krest'yanskogo khozyaistva* [The Influence of Fodder-grass Cultivation on Particular Aspects of the Peasant Economy], Volume 3, Moscow, 1913.

9
Dairy Farming Reprocessing and Cattle-rearing Co-operatives

Associations of peasant households for the purpose of dairy farming are the oldest form of co-operative, going back, so European writers assure us, almost to the fourteenth century. At all events, dairy farming partnerships appear to us to be the most well established, the most highly evolved and, we would say, the most classical form of co-operative organization among peasant households.

The economic problems that have confronted dairy farming co-operatives are of an extremely simple kind; their success has been obvious, their organizational forms have been clearly crystallized and their experience has been accumulated and systematized. All this enables us to describe them in greater detail and to use them as a typical model for the analysis of peasant producer co-operatives in general.

Our basic premise is that a peasant co-operative and its economic activity represent no more than a part of the economic activity of its members – a part which has been detached from the general organizational plan of the agricultural economy and socialized in the form of a co-operative enterprise, while nevertheless inseparably connected with the remaining sectors of the peasant economy. Therefore, when we talk about 'dairy farming co-operation', we use this concept to refer not only to co-operative factories engaged in the manufacture of butter and cheese or to centres for cream production. We use the concept to refer to the whole system of co-operative dairy farming, starting with the stalls of co-operative members and ending with co-operative equipment.

It is, therefore, natural to begin our analysis by examining the economic foundation on which peasant dairy farming co-operation is based, namely, the peasant family farm itself which produces milk as a commodity.

Investigations of the budgets of peasant family farms provide us, in this respect, with a good deal of material. By studying the columns of figures in the statistical records for Vologda and Novgorod, it is possible, as shown in Table 26, to arrive at the following statistical picture of those peasant households which form the basis of dairy farming co-operation.

Table 26

	Vologda province	Novgorod province
Number of members per family	5.3	6.9
Number of workers in family	2.98	3.5
Area under cultivation in the household	2.46	3.37
Head of cattle	3.58	4.7
Including:		
cows	2.09	2.42
calves	1.22	9.21
Gross income of household	448.10	618.08
Including income from:		
cattle-rearing	110.25	99.85
dairy farming	62.71	63.90
from the sale of dairy products	44.66	15.85
Money income of household excluding income from cottage industries	94.78	63.35

Sources: *Materialy dlya otsenki zemel' Vologdskoi gubernii* ('Material relating to land valuations in the Vologda province'), Vol. II, Vologda, 1907; *Byudzhety krest'yantskikh khozyaistv Novgorodskoi gubernii* ('Peasant household budgets in the Novgorod province'), Novgorod, 1918.

Thus in an average peasant household in Vologda, which owns 3.58 head of cattle, 24.6 per cent of overall gross income comes from the products of cattle-rearing and 14 per cent from the products of dairy farming; while 71.3 per cent of all dairy farming products are sold on the market. Receipts from the sale of dairy farming products make up 22.4 per cent of all money income and are second in importance to income derived from cottage industries and trades

Table 27 *Turnover of valuable resources in cattle-rearing for milk production in average peasant homesteads in the Vologda district*

Area sown by households (desyatiny)	Number of households surveyed	Value of cattle at start of year	EXPENDITURE					Total Expenses
			Cattle bought	Labour expended	Cost of fodder	Paid to herdsmen for pasture and other expenses	Maintenance of herdsmen and other general expenses	
0–1.0	14	28.0	–	10.8	15.7	0.8	2.6	57.9
1.1–2.0	47	56.4	2.6	16.3	35.2	2.2	5.1	117.8
2.1–3.0	42	70.6	3.9	19.8	52.4	3.1	6.6	156.4
3.1–4.0	20	98.1	6.4	23.1	57.6	3.4	9.0	197.6
4.5–6.0	9	103.5	–	25.5	72.6	4.5	9.3	220.4
6.5 and above	4	224.3	24.4	37.2	138.3	4.1	20.6	448.9
Average	136	73.1	3.8	19.4	47.3	2.6	6.7	153.0

Area sown by households (desyatiny)	Value of cattle at end of year	Cattle sold	Manure produced	RECEIPTS Hides produced	Meat produced	Dairy products consumed	Dairy products sold	Total Receipts
0–1.0	26.6	2.5	3.3	0.2	0.4	10.5	19.2	62.3
1.1–2.0	57.2	9.2	6.9	0.7	2.5	17.7	37.0	131.2
2.1–3.0	71.7	12.3	9.5	0.9	2.7	18.2	44.2	158.9
3.1–4.0	105.0	8.4	10.8	1.4	4.8	23.6	51.6	210.1
4.1–6.0	101.1	17.7	14.3	2.2	7.1	15.3	63.6	221.3
6.1 and above	225.2	16.5	30.4	3.1	24.2	26.0	123.2	448.6
Average	73.3	10.1	9.1	1.0	3.7	18.0	44.7	159.9

which makes up 50.1 per cent of money receipts. However, the proceeds from the sale of dairy products make up 47.1 per cent of money income derived from the sale of agricultural products.

Thus peasant households, which constitute the foundation of dairy farming co-operatives, are not households that specialize in the production of dairy farming products. Dairy farming accounts for only a small part of the economic activity of the peasant family. It is an activity that takes second place to agriculture and non-farming trades. Its main importance is that it is, by its nature, market-oriented. 71.3 per cent of all dairy farming products are sold on the market, whereas only 22.5 per cent of field products are disposed of in this way.

The money-oriented nature of dairy farming makes it extremely important as a source of money income.

We can get some idea of the way milk production is organized in peasant households from Table 27, which shows the turnover of peasant homesteads of differing sizes.

We can thus see that after deducting the cost of cattle from the debit and credit entries, the expenditure on cattle-rearing for the production of milk is mostly related to fodder (59.3 per cent); and to labour (24.3 per cent). 72.3 per cent of receipts from cattle-rearing come from the sale of milk and dairy products and nearly three-quarters (71.3 per cent) of all dairy products are sold. If we measure the quantities of milk sold in terms of poods [1 pood equals approximately 36 lb] we can calculate the quantities of marketable milk per average peasant household, as shown in Table 28.

As we have already explained, the main importance of dairy products for this economic organization is precisely the fact that these products are marketable. Therefore, the success of this sector

Table 28

Households with a sown area of:	Desyatiny	Poods
	0–1.0	37.0
	1.1–2.0	78.6
	2.1–3.0	97.9
	3.1–4.0	124.0
	4.1–6.0	151.0
	Over 6.1	296.8

[Note: One *desyatina* equals 2.7 acres; one pood equals 36 lb avoirdupois, approx.]

of the economy largely depends on the way marketing is organized and on the prices at which the dairy products can be sold.

If a large consumers' centre, such as a town or factory, exists in the immediate vicinity, the peasant producer can establish direct contact with the ultimate consumer and can organize the marketing of his product directly, without the services of middlemen. However, the geographical area for such marketing is extremely limited. The radius was estimated to be 10–15 versts [approximately 6½–10 miles] in a study of the milk market near Moscow which was carried out in 1911–12 by the Moscow Area Land Board, whose findings we shall utilize later on.[1] Even within this area, however, the direct marketing of milk takes on average about six hours of the peasant's time on every occasion that he undertakes such marketing. It can easily be understood that this outlay only makes sense when the household has a sufficient quantity of milk which is saleable. For a small household with one or two cows, such direct marketing is only possible where there is a town in the immediate vicinity, or where the milk produced over two or three days is all sold at the same time, or, lastly, where the transport of the milk is incidental to journeys made for other reasons.

Table 29 *Percentage of households which market their milk directly and through middlemen*

	Directly	Through middlemen	Mixed	Total
	Percentages			
Area closest to Moscow (1)				
Households which keep a cow for only part of the year	16.7	83.3	0.0	100.0
Households with one cow	33.3	63.6	2.1	100.0
Households with two cows	54.9	39.4	5.7	100.0
Households with three or more cows	71.8	23.1	5.1	100.0
Remote area (4a)				
Households which keep a cow for only part of the year	0.0	100.0	0.0	100.0
Households with one cow	2.0	96.3	1.7	100.0
Households with two cows	8.7	83.1	8.2	100.0
Households with three or more cows	19.5	73.2	7.3	100.0

Therefore, households which want to sell their milk but which do not possess it in large, marketable, quantities, have to rely on the services of middlemen, even within an area of 10 versts [6.6 miles] from the nearest town, as can be seen from Table 29.

We can see that, in a relatively remote area, direct contact with the consumers is maintained only by an extremely small group of households with a large number of cows (19.1 per cent of households with three cows maintain such direct contact). However, all the remaining households have already decided to sell to dealers. This, according to the findings of the study, takes about 0.55 hours, as against more than 6.6 hours which are spent when making direct deliveries to Moscow by horse transport. However, dealers naturally offer a lower payment for milk, since the price in Moscow is about 80 kopecks per pail [*vedro*: equal to approximately 21 pints), whereas dealers in remote areas offer 40–50 kopecks per pail.

Thus a journey of 6 to 7 hours involving human labour and the use of a horse is remunerated, according to the amount of milk sold, as follows:

Table 30

Half a pail	(5 mugs)	17.5 kopecks
1 pail	(10 mugs)	35.0 "
1½ pails	(15 mugs)	52.0 "
2 pails	(20 mugs)	70.0 "
3 pails	(30 mugs)	105.0 "
4 pails	(40 mugs)	140.0 "

[Note: a mug (*kruzhka*) is a standard measure equal to approximately 2.1 pints.]

It can be seen from looking at this table that if we compare the remuneration for delivering milk with the normal earnings from the use of a horse and its driver, there may be a certain advantage in direct deliveries of milk of 1½ pails or more – that is, of amounts produced by farmers with a large number of cows.

So far as areas outside the 10–15 verst radius are concerned, it can easily be understood that given the distance from the town, the transport of milk can be financially profitable only by deliveries of milk in larger quantities than any peasant household can provide. Therefore, in the area outside the 10–15 verst radius, the milk market becomes the kingdom of the middlemen who must, if they are to be sure of a definite income, accumulate a very substantial amount of milk.

Dairy Farming Reprocessing and Cattle-rearing Co-operatives

As we have already seen, direct marketing is virtually impossible for a peasant who has no more than two cows. Only where neighbours pool the milk which they have produced, so as to make up substantial consignments, will it be possible for producers to organize the direct marketing of milk on co-operative principles.

Associations which sought to replace the buyers dealing in cream arose and still exist in the outlying areas of large towns. A survey in Moscow noted the existence of 22 associations for marketing cream; they collected an annual average of 5,339 pails of milk, not counting other products with which they dealt; and the amounts which they collected varied from 363 to 14,675 pails for each association. Despite the fact that nearly all associations are joined into a central association, the co-operative apparatus has been waging an energetic struggle against dealers. It has so far been unable to crush them since it is precluded, as a matter of principle, from resorting to the methods used by cream dealers (for example, giving false measures, adulteration, deception, and so on).

There comes a point when the marketing centre is so far away that neither dealers nor co-operatives can market their milk while it is still fresh, because of the rising cost of railway transport and of transport to the railway station and the significantly higher proportion of the product which gets spoiled in transit. When these distances

Table 31 *Prices paid for fresh milk and for milk in the form of butter, depending on the distance from the marketing town*

Distance from town		Price paid per pood of milk for:	
		Fresh milk	Milk in the form of butter
0	versts	80	52
5	"	70	51
10	"	65	50
15	"	60	49
20	"	55	48
25	"	50	47
30	"	45	46
35	"	40	45
40	"	35	44
45	"	30	43
50	"	25	42
55	"	20	41

are involved, the reprocessing of milk into butter and the marketing of the product in this more easily transportable form – which previously produced low remuneration – now begins to bring a return higher than the return from the sale of fresh milk. Table 31 illustrates in an approximate but visible manner this change in the balance of profitability.

When making the transition to marketing combined with reprocessing, the need to have a broad dairy farming base becomes particularly obvious, since this form of marketing is possible only when the milk is reprocessed mechanically, owing to the decisive advantages of mechanical over manual reprocessing. But the profitable use of the equipment of a butter-producing factory is possible only when the amount of milk reprocessed is substantial. Otherwise, the costs of amortization and the cost of the foreman's wages will be disproportionately high in relation to each pail of reprocessed milk. Therefore, for a small peasant household like the one in Vologda, which we examined above, the organization of mechanical reprocessing and marketing is only possible as part of a co-operative system including many other households.

Practice has shown that the use of mechanical cream-producing equipment can be profitable where at least 4,000–5,000 poods of milk are reprocessed every year. Let us, therefore, try to calculate how many peasant households need to combine in order to create such a dairy farming base.

If, instead of taking averages, we look at the figure for milk production in particular households, we get the data shown in Table 32, by examining the budgets of one in ten of the Vologda households.

As can be seen from Table 32, some households do not sell milk at all. Among the Vologda households these constitute 19.2 per cent of the households whose budgets have been analysed.

If we calculate that the average delivery of milk to the market by a household which sells milk is 121.4 poods, we can then reckon that the profitable use of mechanical equipment for marketing purposes requires at least 40 households which share in its use. Practical experience gives us figures which are quite close to this. Thus an average partnership in Novgorod has 60 members who own 140 cows and produce 5,600 poods of reprocessed milk; and an average partnership in Kostroma has 50–60 members owning 110–50 cows and producing 5,000 or 6,000 poods of reprocessed milk. Partnerships in Vologda exist on an even greater scale: they include, on average, 144 members owning 289 cows and they produce 18,841 poods of reprocessed milk every year.

This, then, is the essential foundation for the setting up of a

Dairy Farming Reprocessing and Cattle-rearing Co-operatives 171

Table 32 *Amount of milk sold by households annually*

No. of households according to Vologda statistical records	Milk sales (poods)	Revenue from milk sales (roubles)
10	58.5	25.20
20	69.5	29.90
30	No sales	—
40	No sales	—
50	54.0	23.20
60	97.7	44.10
70	No sales	—
80	125.0	54.00
90	80.0	34.58
100	185.0	80.00
110	146.5	63.20
120	174.5	75.25
130	130.0	56.00
136	415.0	178.88
Average for 136 households	68.8	44.30
Average for 110 households selling milk	121.4	53.5

butter-producing association. It has to be noted that an average household, which belongs to a butter-producing co-operative and which has two cows and supplies 130.9 poods of milk to the market, is slightly bigger than an average peasant household in Vologda and approximates more closely to households of the category which sows between 3.1 and 4 *desyatiny* [i.e. between 8.37 and 10.8 acres] of land.

Thus 144 peasant households, when organizing the marketing of their dairy farming products, pool a significant amount of the milk which they produce; and they arrange for it to be reprocessed by a co-operative for the purpose of jointly marketing the butter which constitutes the final product.

Let us now examine how the co-operative enterprise is built up on this foundation. For the purpose of building it up, there are four organizational problems which have to be solved, namely:

1. The acceptance of the product as fit for marketing;
2. The organization of its reprocessing;
3. The organization of its marketing; and
4. Financial settlements.

Let us examine each of these problems in turn.

The organization of the formal acceptance of the product relates to the organization of milk production in the members' households. When they organize their dairy farming in co-operative forms for marketing purposes in order to earn a financial return, peasant households nevertheless continue to carry on dairy farming for the purposes of their own consumption; and their consumption requirements are sometimes interwoven with the requirements of farming for the market and may take precedence over the latter. On the other hand, dairy farming may be linked with agriculture as well as other predominant economic activities on the farm; with the result that dairy farming is affected by the way these other activities are organized.

Thus, for example, in an area near Moscow, we have the following circulation of resources on a dairy farm during different seasons of the year.

Table 33 *The influence of the seasons on dairy farming resources*

Season	No. of cows per home selling milk	Daily supplies of milk (pails)			Sales as percentage of milk-yields	Prices per pail (kopecks)
		Milk-yields	Sales	Residue		
Period of large milk-yields in summer	2.43	2.02	1.43	0.68	66.1	77.3
Period of small milk-yields in summer	2.30	1.32	1.01	1.31	76.6	81.1
January–March 22–29	2.39	0.95	0.88	0.07	93.5	106.1
April (Fomina nedelya)	2.10	1.49	1.11	0.38	74.6	87.6

Dairy Farming Reprocessing and Cattle-rearing Co-operatives 173

Looking through Table 29 we can see that in this area near Moscow, which might appear to be totally governed by market forces, the amounts of milk produced are, in fact, determined not by demand or by price levels, but by spring calving and by the availability of fodder. It is only the high prices in winter which lead the peasant households to increase the number of cows they keep and to curtail their own consumption.

A cream-producing association built on this kind of economic foundation will have a very unstable basis as an enterprise, since it will suffer from a milk shortage at the times when the demand for milk is greatest, while it will be saturated with milk during periods when storage and preservation are extremely difficult. It is true that these fluctuations are less severe in the case of an association which reprocesses milk into butter, which is a less perishable product. But even so, the fluctuations make themselves felt. Therefore, all co-operatives, acting on the basis of their interests as enterprises, always try for as long as possible to prevent the milk which they collect from being spoiled. And they do so by trying to persuade their members to change from spring to autumn calving and by making their members undertake to deliver at least a stipulated minimum quantity of milk so as to guarantee an adequate and permanent supply.

The co-operative reprocessing of milk is, like any other co-operative undertaking, carried out for the ultimate purpose of increasing the remuneration of the agricultural labour applied to the reprocessed product. The successful achievement of this purpose with the concerted backing of workers in butter-producing co-operatives depends on the following conditions, when reprocessing is carried out:

1. The amount of expenditure on production;
2. The technical standard of manufacture;
3. The manufacture of products of high quality;
4. The quality of the material;
5. The rational utilization of skim milk and butter-milk; and
6. The level of the market prices of the raw material and of the final product.

We shall examine each of these conditions on the basis of a detailed study of butter-producing partnerships in Vologda, which was presented to the Vologda co-operative congress of 1913. According to the data assembled by this study, the cost of producing butter is made up of the following components:

174 *Dairy Farming Reprocessing and Cattle-rearing Co-operatives*

1. The wages of the workers and foreman;
2. Payments to those who transport the milk;
3. Payments to clerical staff;
4. Management;
5. Transport of the butter;
6. Amortization of stock and property;
7. Raw materials;
8. Miscellaneous expenditure.

Needless to say, the larger the partnership, the greater its overall total of expenditure; and if we calculate it *per unit of reprocessed milk*, we can observe great variations between individual cases. The cost of reprocessing varies between 5.8 and 18.2 kopecks per pood of milk. This enormous difference may be due to the distance between the place where the partnership works and the market-place; or it may be due to the partnership's technical standards of work or to the size of the partnership. However, the size of the partnership is less important than might be supposed.

When comparing the cost of equipment with the capital and equipment owned by a peasant household, we arrive at the following outline picture of a butter-producing co-operative in the Vologda province:

1. It is based on 144 peasant households.
2. Their overall capital comprises 115,000 roubles.
3. This includes capital consisting of 289 head of cattle (289 cows) worth 14,450 roubles.
4. The factory equipment is worth 2,000 roubles.
5. The factory equipment comprises:
   (a) 1.8 per cent of the capital of the peasant households; and
   (b) 13.8 per cent of the value of their cattle.
6. The overall gross income of the 144 households equals 64,500 roubles.
7. This includes the value of milk amounting to 12,000 roubles.
8. The value of the milk handed over to the factory (19,000 poods) equals 8,000 roubles.
9. The cost of reprocessing is 1,800 roubles.
10. Receipts from the sale of butter amount to 11,000 roubles.
11. Proceeds from the sale of waste material equal 1,100 roubles.
12. Payment to the butter-producing partnership for one pood of milk is 54 kopecks.
13. Milk used for reprocessing amounts to 13.3 per cent of the household's income.
14. The profit from butter production represents:

(a) 21.2 per cent of the value of the milk; and
(b) 2.8 per cent of gross income.

Such are the typical entities producing most of the butter which goes on to the world market. They are partly based on individual households and are partly socialized in a co-operative form. Every success or failure in the co-operative part of the system has an immediate effect on the family farms which form its base. Conversely, every shortcoming in the family farms undermines the stability of the co-operative part of the system.

When examining the conditions for the success of butter manufacture, we have more than once had occasion to note the influence which the economic activity of the membership exerts on the economic activity of the co-operative. And conversely, a number of recent investigations have demonstrably shown the organizing influence of co-operatives on their members households. Thus researchers on co-operatives in Vladimir write that:

> When we compare the present state of the farms with the state of these farms 15 years ago, we are unable to discern any significant increase in the use of intensive methods in the farms of today. However, there have been significant changes in the way the farms are organized, especially with regard to tillage. Fodder-grass cultivation is beginning to play a prominent role; and there is a significant expansion of the use of labour-intensive crops.

The changes in tillage which have taken place on our farms over a period of 15 years are shown in Table 34. In these cases, no radical change in tillage methods is yet apparent. What we find here is the old three-field system, in which two fields are set aside as a special area for clover, which does not form part of the general seed-turnover. All the same, this does represent a step forward. An attempt is being made to add a new ingredient to the overall resources of productive farming – namely, a new kind of fodder. The substantial changes which have occurred in these farms during 15 years have meant that the farms are better provided with cattle; and that cattle have become distributed more equally among different farms.

The existence of a larger number of cattle is now making it possible to dispense with outside help in working the land and thus to offer more jobs on farms to those who had previously sought work outside agriculture. At the present time we can observe a greater use of intensive methods of tillage as well as the use of new kinds of crops.

Table 34 *Proportion of field crops*

Year	Crop	Spring crops						Grasslands sown		Total
	Rye	Oats	Buckwheat	Flax	Potatoes	Peas	Total	Wine	Clover	
1893	47.5	18.0	24.1	5.5	4.6	0.3	52.5	—	—	100.0
1913	39.7	4.4	22.1	10.2	9.8	—	49.6	3.1	10.7	100.0
Increase (+) or decrease (−)	−7.8	−13.6	−2.0	+4.7	+5.2	−0.3	−2.9	+3.1	+10.7	0

Dairy Farming Reprocessing and Cattle-rearing Co-operatives

We have so far been examining the organization of the butter manufacture co-operative system and the organization of the collection and reprocessing of milk. We should not forget, however, that, here as elsewhere, the main factor in the success of the work of a co-operative apparatus is the marketing of the finished product.

Marketing conditions as well as marketing policy are what determine both the forms and the scale of the co-operative organization itself. It is precisely in relation to dairy farming co-operatives that the influence of these factors is particularly important because, in this case, milk can be marketed not only in the form of various kinds of butter but also in the form of fresh milk or various kinds of cheeses or other dairy products. The eventual revenue from the sale of milk depends, in the last resort, on the ability of those who organize it to study and exploit the complex situation on the dairy market by making appropriate adjustments to their economic apparatus.

What is particularly complex is the building of market relationships in the market for fresh milk in the outlying areas of towns, where seasonal price fluctuations are sometimes exceptionally severe, as can be seen from Figure 11.

The price of butter, like that of milk, although to a lesser extent, is itself subject to seasonal fluctuations. Thus, for example, variations in the price of butter and in the corresponding prices of a pood of milk in the Vologda province showed this picture for 1912 (Table 35).

Table 35

	Price of 1 pood of butter	*Proceeds of sale per pood of milk*	*Difference in payment for milk compared with payment in April*
January	15.52	68.9	14.4
February	14.41	61.7	7.2
March	13.25	57.1	2.6
April	12.82	54.5	—
May	13.34	58.6	4.1
June	13.56	58.5	4.0
July	14.91	64.9	10.4
August	16.81	76.9	22.0
September	17.69	80.8	26.3
October	19.75	93.1	38.6
November	19.75	102.5	48.0
December	22.53	87.0	32.5

178 *Dairy Farming Reprocessing and Cattle-rearing Co-operatives*

Figure 11: Price of a mug (half a litre) of milk in 1924 during diffferent months

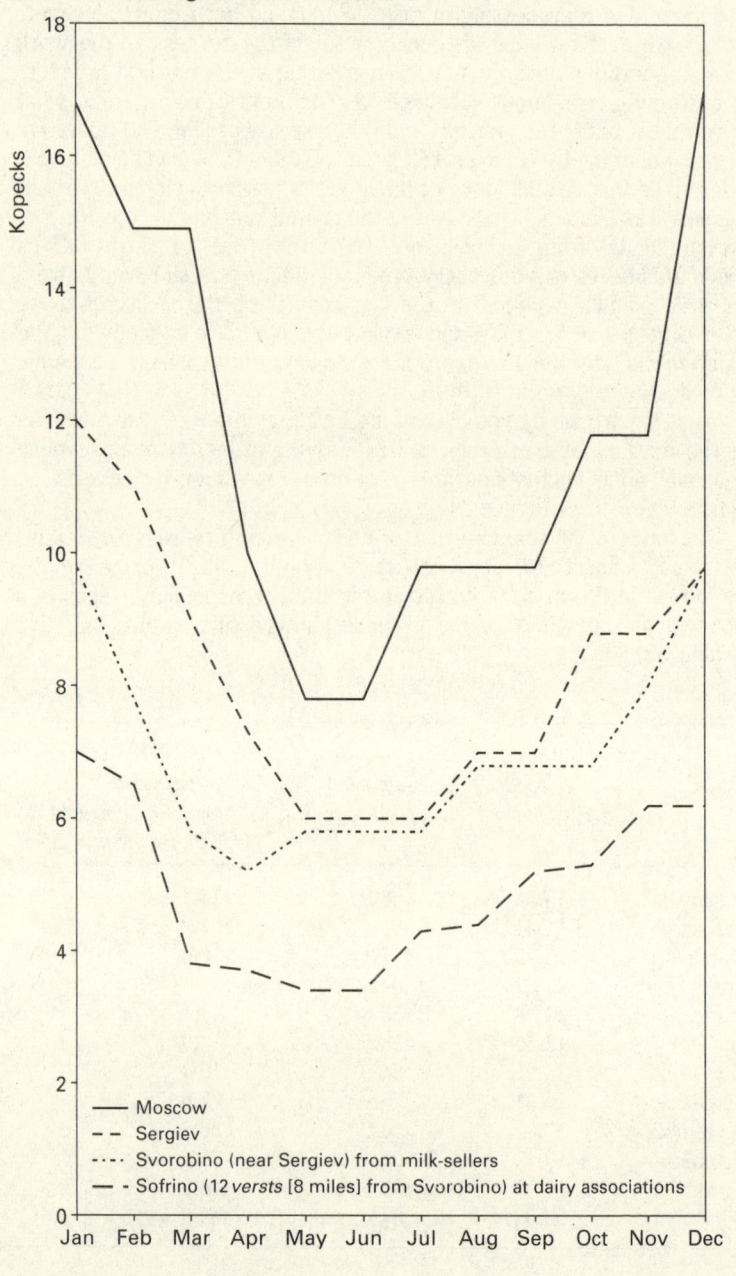

— Moscow
-- Sergiev
··· Svorobino (near Sergiev) from milk-sellers
— - Sofrino (12 *versts* [8 miles] from Svorobino) at dairy associations

Dairy Farming Reprocessing and Cattle-rearing Co-operatives 179

If we compare the lowest payment for milk, in April, when a partnership was able to pay a total of no more than 54.5 kopecks per pood, with the payment of 102.5 kopecks in November, we can see a difference in the payment for the product equal to 48 kopecks, that is 87.2 per cent.

Financial calculations alone will lead co-operatives to sell all their butter during the autumn and winter when prices are high; and, during the remainder of the year, to store their supplies of unsalted butter in well-equipped ice-boxes. For this reason, the problem of installing refrigerators in central co-operative organizations is one of the most immediate problems facing Russian butter producers, since, if it is successfully resolved, it can achieve an increase of 20 or 30 per cent in the peasants' revenue from their milk.

There is, however, another solution. We have already noted that in addition to reprocessing milk into butter, a dairy partnership can also undertake the manufacture of cheeses. The prices of cheese have been noted for their great stability. The price paid for milk reprocessed into cheese is significantly below that paid for butter products in November and December; but, on the other hand, the price paid for cheese was higher than that paid for butter during the spring and summer. This can be seen from Figure 12. Therefore,

Figure 12: Seasonal prices of cheese and butter during different months

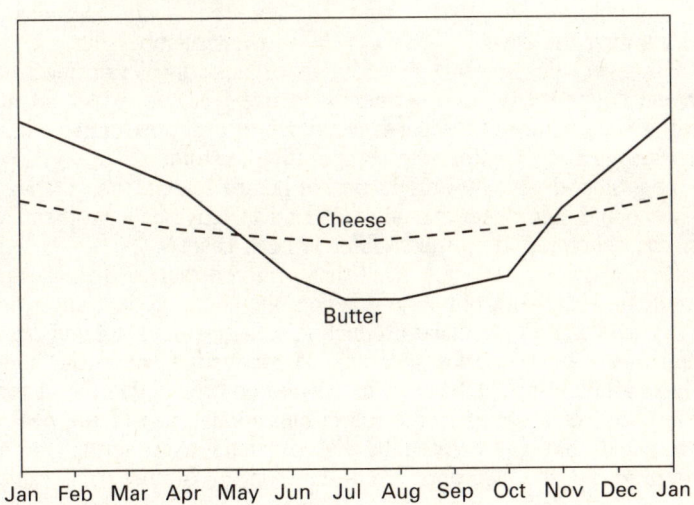

although it is impossible to organize a perfect system of storage in refrigerators, it is nevertheless possible to divide output according to two periods, namely (1) the period from February to July, which is taken up with the production of cheeses; and (2) the period from August to January, which is occupied with butter manufacture.

The changeover to cheese-making is convenient in this case for the further reason that cheese, which is an item of mass consumption in Western Europe and America, has a considerably wider market than butter and, therefore, can always be sold more easily than butter. It is true that cheese-making involves a much slower capital turnover and that a partnership engaged in cheese-making needs considerable resources in order to store the cheeses while they are maturing. But if well-organized co-operative credit facilities exist, this is no longer such a serious obstacle as it was in the past.

★   ★   ★

Potato-grinding, vegetable-drying and other co-operatives engaged in primary reprocessing are essentially almost identical to the butter manufacture co-operative system which we have just analysed. While leaving agricultural production under the control of individual peasant households, these co-operatives socialize all the work of reprocessing; and they organize the marketing of the final products. When we examine their organizational principles stage by stage, we can see the same methods of delivering raw materials, the same depersonalization of the raw material when it is reprocessed, the same methods of marketing and, lastly, we can see the same methods of distributing the revenue amongst the peasant membership.

However, despite the similarity of organizational principles, we should nevertheless note certain distinctive features, which are due to the very nature of the particular types of economic activity which are being brought within the co-operative system.

Vegetable-drying, the production of potato starch, the growing of tobacco and sugar beet and all other similar activities have their own special characteristics, which differ in each case.

It should be noted first of all that – in contrast to dairy farming products – the work of potato-grinding, vegetable-drying, sugar beet cultivation and other similar production activities does not and cannot offer any wide scope for spontaneous production within the home. These kinds of production can be carried on only within the factory. They are not a derivative of the organizational plan of the peasant household, but are created by the peasants from scratch or are captured by the peasants from the control of industrial capital.

Therefore, they are not in competition with the individual peasant household.

On the other hand, in the overwhelming majority of districts, this kind of agricultural base for co-operative industry – that is, the growing of sugar beet and the industrial cultivation of potatoes and vegetables – is itself possible if, and only if, reprocessing work organized on factory principles is carried on in the vicinity. For in this case there can be no substantial local market for the reprocessed raw material; and the raw material is unsuitable for transportation to the larger markets owing to the very considerable weight per unit of the commodity.

Therefore, the existence of an agricultural base for these kinds of reprocessing is itself impossible without co-operative reprocessing in the factory, and vice versa. Wherever potato-grinding or vegetable-drying co-operatives appear, they will give rise to potato-growing and market gardening organized on industrial lines, which would previously have been unthinkable. It is therefore impossible to examine the profitability or otherwise of these co-operative undertakings without at the same time examining the profitability or otherwise of potato-growing or market gardening. Cases may possibly arise where vegetable-drying or starch production make losses if taken on their own and therefore cannot, for example, exist in a capitalist form; but where their losses will nevertheless be offset by the benefits which these crops provide for agricultural production itself.

Cases of this kind may arise in densely populated areas which suffer from a shortage of land and, consequently, from an underemployment of manpower. In such cases, the appearance of labour-intensive crops such as potatoes or vegetables will always provide employment for previously redundant manpower and thus increase the labour earnings of the peasants throughout the whole area. In this case the peasantry will stick to the labour-intensive cultivation of potatoes and vegetables and to the co-operative system needed for this purpose, even if, from a book-keeping point of view, this makes a loss. The need for, and profitability of, a co-operative system will be determined in this case not by the conditions obtaining in the co-operative enterprise but by the existence of its agricultural base.

Another way in which the types of co-operation which we are examining differ from those of butter manufacture is that they find it possible and profitable to use large-scale factory installations. As a result of this, co-operatives, particularly those engaged in sugar beet production, must, when they start their business, possess a considerable basic capital together with equipment on a scale which, if it to be fully utilized, requires a quantity of raw material considerably greater than the raw material stocks owned by the co-operative's founding members. This heavy dependence on capital considerably weakens the standing of a co-operative *vis-à-vis* a capitalist enterprise; and it hampers the rapid and easy development of co-operative undertakings.

A good many of the distinctive features of the above cases are of a purely technical nature. The most important thing here is that, unlike milk which is sent for reprocessing every day throughout the year, potatoes and vegetables are produced on the farm all at the same time, after the harvest has been gathered. This in turn raises the problem of the organization of storage, which is sometimes extremely complex as well as being very important in view of the possibility of spoiling.

It is also possible to undertake numerous other kinds of reprocessing in addition to those just enumerated.

NOTES

1. N. Makarov, P. Kolokol'nikov and others, *Molochnoye khozyaistvo v Moskovskom uyezde* (Dairy Farming in the Moscow District), Moscow, 1914.

10

Peasant Co-operation for the Purpose of Cattle-rearing

If we open the pages of Dr Eduard David's well-known work on *Socialism and Agriculture*, or if we consult books by other authors who dealt in their time with the struggle between large-scale and small-scale agriculture, we find the most sincere assurances to the effect that of all branches of agriculture, it is precisely modern intensive cattle-rearing that is, *par excellence*, the province of small-scale production.

Modern cattle-rearing consists of complex biological processes whereby a living organism converts the crude raw material of fodder into food products containing an unusually high concentration of assimilable energy. The organization of this kind of cattle-rearing is both unique and non-mechanical. Its processes are so completely individualized, and they require so much assiduous attention in the care of each animal, that cattle-rearing, so David insists, can be properly undertaken only by a small farmer who has a personal interest in the success of what he is doing and who can constantly observe the small number of animals which he keeps.

A peasant family that keeps three to five head of cattle is alone capable of putting in the effort that is essential to the animals' care. Such effort cannot be provided even by dozens of workers, who merely carry out general routine directives from the management, while having little personal interest in the success of their work. David ended his argument by saying:

> If one bears in mind the enormous importance of assiduous care for the lives of the ennobled species, one can easily understand why practically all those who are competent to judge recognize that in relation to cattle-rearing, a small-scale

farm directly managed by its owner enjoys an immense advantage over a large-scale farm.

Despite this, however, the practice of cattle-rearing in the agriculturally advanced countries of Western Europe has shown that, even in this area, where the peasant farm has had a special importance, such a farm is still able, without sacrificing its individuality, to single out certain types of activity, which have been shown to be more profitably organized as large-scale farming. Indeed, when we look at the organization of peasant cattle-rearing, we find a good many cases where the economic measures which are required are beyond the usual capacities of the small self-employed peasant farm.

Thus, for example, as regards the very foundation of cattle-rearing, that is, the reproduction of the animals, a small farmer with only one or two cows and a horse is totally helpless, since he cannot, with such a herd, exploit even a quarter of the animals' breeding capacity. It makes no sense for a peasant family to keep a bull or a stallion on the farm for the purpose of breeding, since the costs of keeping such an animal – in relation to two or three cows instead of twenty or thirty – will be disproportionately high and will not be compensated by the increase of cattle. Therefore, the collective use of studs for breeding has been a fairly long-standing practice in the peasant economy. In areas of communal land tenure, this has long since taken the form of unrestricted mating between a jointly-used herd and a jointly-used bull.

But in areas where land has been owned by separate households, a widespread practice developed, on a commercial basis, of keeping a bull [and offering its services as a stud] as a method of paying for cows purchased from others. The struggle against entrepreneurs of this kind, who were often selfish and unscrupulous in exploiting the owners of cows, resulted in the joint co-operative acquisition and upkeep of animals for stock-breeding purposes.

However, in the case of a peasant household which has set out along the path of agricultural progress, the question of the pedigree qualities of its cattle's offspring has a relevance which goes beyond the maintenance of the number of cattle. It gradually becomes a problem of constantly improving the productive qualities of this cattle. The need to improve the quality of the cattle becomes increasingly important and grows especially acute with the development of the production of dairy products and other products of cattle-rearing for sale on the market. Industrial cattle-breeders, who send out most of their output to the market, start to become especially sensitive to the valuation in money terms of the quantity as well as the quality of this output.

The rouble is always the best teacher for anyone learning about weights and measures. The peasant begins, from his everyday experience, to discover how the remuneration for his work depends on the quality of the animals that he keeps. Agronomists in the Vologda province have more than once observed, and not without reason, that the peasant's interest in the improvement of his cattle becomes visibly aroused only with the development of co-operative industrial butter manufacturing. Cattle-breeders, when they cease to be concerned only with consumption and turn to production for the market, will naturally seek to raise quality by improving the pedigree of their cattle.

However, although we can discern a new need within the peasant economy to improve its cattle, we are almost entirely unable to indicate any methods which might help a small family farm to satisfy this need. It is, admittedly, possible when buying cattle to select good specimens; but such a selection can only be made out of the existing stock. If carried out over several decades, it might eventually have an effect on the average stock of cattle. But this method of spontaneous improvement is a slow one, and it is in no way guided by social or agronomical criteria.

But so far as the small, fragmented farms are concerned, all other paths are closed, because artificial selection and extensive cross-breeding require the availability of a great deal of biological material, the setting up of special stud farms, breeding-grounds and tens of thousands of roubles of capital. Most important of all, they require a highly educated staff of agronomists who are versed in the art of cattle-rearing and who have experience in this field. None of this can be attained unless the work is organized on the largest possible scale. And it can be attained by the peasantry only if a co-operative approach is adopted towards the solution of the tasks in question.

Societies of cattle-breeders, combining the efforts of hundreds of individual farms, can gain access to stock-breeding material which is of such a vast quantity that it is beyond comparison with any private cattle-rearing farm. Such associations, if they are provided with adequate resources, can improve the cattle in ways which are more profound and more successful than a private breeding-ground for cattle.

The difficulties facing the peasant are just as serious with regard to the rational organization of the feeding of the cattle, which requires constant laboratory observation, both of the content of the fodder and of the milk-yield. Modern methods of individual feeding, which correlate the fodder given to the cattle with the live weight of the cattle and their daily milk yield, require extremely complex calculations and assessments.

The care of cattle organized on these principles is aimed at getting the highest possible return, in the form of products, from every unit of fodder which is fed to the cattle; and it produces exceptionally good economic results in terms of revenue. But it is possible only with the help of laboratories and the help of specialists who have experience as well as an adequate training. Such things are accessible only to a large-scale farm which has substantial supplies of milk and which can pay the costs of agronomical observations by increasing its production of milk. For a peasant household, which stands isolated and on its own, all these advances in the techniques of agronomy will forever remain a closed book. Even the doubling of the milk yield of a farm with two cows would still be insufficient to pay a month's salary for a specialist in agronomy. However, exactly the same kind of monitoring by agronomists of the ingredients of the fodder given to cattle, as well as every kind of laboratory research, can easily be undertaken for a group of even the smallest farms – provided that the total number of cattle which they keep is large enough to ensure that such measures are profitable and can adequately pay for themselves through their economic results. Associations which handle this are widespread in the West and are successfully tackling the problems mentioned.

The co-operative principle has been bringing the latest advances in the science of agronomy into the stalls of peasant households, and has been producing an immediate and tangible benefit from these advances. This same co-operative principle has been enabling small farms, which are deprived of pasture and obliged to keep their cattle in stalls, to graze their young animals on good mountain pastures which are jointly leased by co-operative cattle-breeders and which ensure that they have strong and healthy young cattle.

When surveying the examples mentioned above of co-operation in the organization of cattle-rearing, we can divide them into two categories:

1. Co-operation for the improvement of cattle pedigree; and
2. Co-operation for improving the conditions for the exploitation of cattle.

In the USSR none of the different types of co-operation in relation to cattle-rearing has come anywhere near to completing its initial development. Nor have their organizational types yet been firmly established or developed into a final form.

11
Co-operative Insurance

In one of the early chapters of this book, we carefully traced the processes by which capital circulating in the peasant economy is replenished, and we identified the ways in which a co-operative credit system could powerfully assist these processes.

It must be noted, however, that co-operative credit is by no means the only or even the best way of providing co-operative support for the replenishment and maintenance of capital in the peasant economy. In this respect, co-operative insurance is just as important, if not more so.

It is true that the economic problems of co-operative insurance are more complex and are harder to solve than those of small-scale credit. This is the reason why viable solutions were found in practice only a long time after the great principles of Raifeizen were formulated. The development of co-operative insurance is still only in its early stages. Nevertheless, the right path has already been explored and discovered; and the most viable organizational forms have been arrived at by a process of selection. The example of the co-operative insurance of cattle, which has become widespread in Belgium and France, holds out the prospect of rapid development for this new branch of the co-operative movement.

The problem of insurance is at first sight simple. Every farm over the course of the years suffers unexpected losses or accidents which hit part of its turnover capital or basic capital: for example, cattle plague infecting a horse or cow, fire in a house or shed, damage caused by hail, and so on. Although from the point of view of each individual farm, these unforeseen events are both unexpected and fortuitous, nevertheless from the point of view of a very large number of farms which are situated over a substantial land area, they represent a normal phenomenon which recurs from year to year and which, in any year, affects only an insignificant percentage of farms. The larger the number of farms that we survey, the more constant and stable this percentage will be. It is therefore entirely feasible, in respect of any large group, to calculate in advance the amount of loss which will be suffered by individual farms belonging to this group, as the result of a 'fortuitous' disaster. By dividing this amount between the farmers, insurance can provide each of them with an alternative to the risk of heavy losses, through the payment of small contributions into a fund which covers losses suffered by farms in a particular year.

The practice of insurance has shown that in relation to large groups these percentages are so constant that they can be used for making firm calculations and can provide a basis for setting up enormous commercial enterprises. The vast organization of fire insurance and life insurance is an undoubted proof of this. But in the case of some types of economic misfortune, the routine operations of large commercial insurance companies are not a very suitable method for solving the problem of insurance. This applies to types of loss where the real value of the item insured is very variable; or where the loss may have been caused by the malice or negligence of the owner, which is hard to detect.

Insurance of this kind – which mainly relates to the *insurance of cattle* and the *insurance of crops sown* – requires constant and vigilant observation of the item insured. This can be undertaken at a low cost only where the insurance agency is situated in close proximity to the insured and where the insurance itself is based on comradely supervision. Indeed, a detailed examination of co-operative forms of insurance against cattle plague and damage from hail indicates that agricultural co-operative insurance has an advantage over commercial insurance by virtue of being a stable enterprise which incurs no losses. The reason for this is that it is based not on the principle of fixed premia versus fixed payments in the event of accidents, but is essentially based on the comradely apportionment of the victim's losses between all who participate in the insurance.

This last-mentioned fact gives us grounds for hoping that, in the course of time, the co-operative principle will replace the commercial principle in other branches of insurance as well. But for the moment, we can deal at length with only two kinds of co-operative insurance – relating to cattle and crops – which we shall now proceed to describe in greater detail.

THE INSURANCE OF CATTLE

One of the gravest disasters which can afflict a small peasant household is cattle plague among its horses or cows. Assuming an annual money turnover of 150–200 roubles, of which three-quarters is spent on the barest necessities of life, the expenditure of some 50 roubles in order to acquire a new animal is very large; and such a payment can very often spell economic ruin, particularly during the months when the harvest is still a long way off. A household which has lost a horse will very often be totally unable to afford a new one and will get into the distressing position of having no horse at all, from which it may be difficult to escape. One is bound to assume that any peasant householder who has lost his livestock and is having financial difficulty in replacing it would be thankful beyond words if offered the chance of paying for it by instalments, spread, let us say, over five or six years. The whole misfortune is that horses and cows are not sold in the market on the basis of payment by instalments; or if they are, the trader imposes an onerous rate of interest. But it would, however, be possible, to arrange matters between the peasant householders themselves, so as to ensure that money spent on the replacement of losses of cattle is repaid by instalments spread, not over two years, but over a period of twenty years. It is this possibility which is realized through the insurance of cattle.

It may confidently be said that a peasant household with one horse is certain during the course of twenty years to experience at least one case of unexpected cattle plague. The peasant householder who foresees this will insure his horse in just the same way as he insures his house and his shed; and he will make an annual payment of, let us say, 2 roubles. In the event of an occurrence of cattle plague, he will receive 40 roubles compensation from the insurance agency; and out of this sum he will be able to replace the animal which has been lost.

The only question that arises here is, in effect, whether the insurance of cattle can be organized on the same principles as the insurance of buildings against fire, and how this insurance ought to be undertaken. To answer this question we need carefully to examine the reasons for outbreaks of cattle plague, as well as their frequency in relation to the total numbers of livestock. In doing this we need, first of all, to distinguish as sharply as possible between two kinds of death rate, namely:

1. *The ordinary death rate*, caused by ageing, accidents, lightning, wild animals or various not very contagious diseases; and

2. *The death rate from epidemics*, that is, deaths due to epidemic diseases (epidemic pneumonia, tuberculosis, malignant anthrax, and so forth).

More than three-quarters of the cases of cattle death here in Russia can be attributed to the ordinary death rate. Epidemic diseases become increasingly rare as cattle-rearing improves and as the country becomes more cultured. Therefore, the main task of this study is to examine the nature of this ordinary death rate.

Within any large herd – comprising, let us say, 200 head of cattle – between five and eight animals will die every year through chance, or as the result of ordinary causes of whatever kind. In large, well-run farms, this ordinary incidence of cattle death is always allowed for in advance, and a certain sum of money is set aside, in the farm's financial estimates, for the replacement of losses of cattle. It comprises a small percentage of the total value of the cattle (usually no more than 4 per cent) and is included in the expenditure on the maintenance of the cattle.

Let us now suppose that one such large herd is bought up by peasants and therefore passes into the possession of small-scale owners, each of whom owns one, two or at the most three cows. The incidence of ordinary cattle death remains as before; and it can be anticipated with equal certainty that five to eight head of cattle will be lost each year through the normal death rate. However, it is totally impossible to know in advance exactly which of the animals will fall victim to such accidents. Therefore, unlike a large-scale owner, the owner of, let us say, two cows, which originally belonged to this herd, cannot set aside a sum equal to 4 per cent of the value of the herd as a cover against cattle death since it is quite possible that over the course of many years he will not have a single case of cattle death, although the other owners will suffer losses of cows. Even if this owner is sufficiently far-sighted to set aside 4 per cent of the value every year in the hope of eventually saving enough to provide for future accidents, his good intentions will not always be realized. Thus, for example, if, in the ordinary course of events, one of his cows dies in the very first year, he will have to pay not 4 per cent of the value of his cattle, but 50 per cent, assuming he owns two cows. And if he has only one cow, he will have to pay 100 per cent.

Thus for a large farm forming part of a landed estate, the ordinary incidence of cattle plague is a normal economic phenomenon and is a minor item of annual expenditure alongside other expenditure on cattle-rearing. But we have already seen that for a peasant household, even assuming this same ordinary incidence of cattle death, the loss of an animal will be an appalling calamity.

It is obvious from this comparison that if any institution could be found which was prepared to collect from our peasant householders an annual sum equal to 4 per cent of the value of their cattle, then the money thus collected could easily cover these losses due to cattle death, just as it does in the case of large farms forming part of landed estates. And the peasants would, by making a comparatively small payment, rid themselves from the serious losses caused by cattle plague, thus preventing all kinds of unforeseen disturbances in the economic equilibrium.

The only prerequisite for success is that such insurance payments should extend to the greatest possible number of livestock. Indeed, if, for example, we take a number of neighbouring villages, we will see that in some villages the incidence of cattle death in the current year was higher than usual, while in other villages it was lower. In the following year the opposite may be true: in the former group of villages the incidence of cattle plague may be slight, while in the latter it may be greater than usual. But if the two groups are treated as a whole, their losses will be mutually offset and the overall incidence of cattle death will be more constant.

The study of the realities of everyday life confirms this conclusion: the larger the numbers of livestock insured, the more constant will be the incidence of cattle death from year to year. Thus in Belgium, for example, the incidence of cattle death in relation to the country's total livestock showed the following extremely stable picture during the most recent years for which we have information (Table 36).

Table 36 *The incidence of cattle plague in Belgium*

Incidence of cattle plague among:	1st year	2nd year	3rd year	4th year	5th year	6th year	7th year
				(Percentages)			
Horses	3.7	3.6	3.7	3.7	3.5	3.7	3.6
Cattle	4.3	4.2	4.4	4.2	4.2	4.3	4.3

The figures are so constant that they can be used as a basis for the most definitive insurance calculations. Exactly the same principle of covering the largest possible numbers provides the basis for all calculations relating to fire insurance, as well as for all other insurance operations.

The situation is entirely different in the case of cattle death caused by epidemic diseases. These do not occur every year; however, when they do occur they usually spread over vast regions and increase the incidence of cattle death two or three times over.

If the agencies that undertake insurance against ordinary, normal losses of cattle were also required to pay in cases of cattle epidemics, then they would be ruined by the very first epidemic. If they were to demand increased insurance contributions, this might, of course, soften the blow for those who had lost their cattle. But it would considerably diminish the value of insurance for cattle-rearing as a whole.

Therefore, insurance against epidemic diseases cannot be undertaken by the methods used for insurance against the ordinary death rate. It is impossible to use the method of spreading out losses from ordinary cattle death over the largest possible number of livestock. What is necessary is to spread out the losses over the greatest possible number of years, so that they can be paid off by instalments. This can be achieved either by means of credit or else by a special fund of reserve capital, of which more will be said below.

Thus our first step must be to treat insurance against epidemic diseases as a separate undertaking and to organize it on special principles. Such are the foundations on which insurance becomes possible. Let us now look at the organizational forms through which this possibility is realized.

Among all the countries of Western Europe, the most highly evolved as well as the simplest and cheapest forms of co-operative insurance are to be found in Belgium and France. Already at the beginning of the nineteenth century, there existed in the heart of the Belgian countryside peasant associations that were free from outside control, were not registered with the authorities and had their own somewhat distinctive arrangements. They had no money capital; but whenever an animal died whose meat was fit for consumption, the members of the association undertook to buy the meat from the owner in proportion to the number of cattle which they had insured, at current market prices. And in the event of the meat being unfit for consumption, the owner was paid by his colleagues the amount of money which they would have had to pay for the meat had it been fit for consumption.

When operations were expanded, societies of this type – which were not very convenient for their members – were superseded by other types of society which built up a certain amount of capital out of annual contributions. This capital was used to pay benefits, without troubling all the members on each occasion. The most usual of these evolving systems was one which required the society to pay, out of its funds, the difference between two-thirds of the value insured and the sum which the owner had been able to fetch by selling the animal's carcass. This system gradually developed into a purely monetary system whereby an owner immediately received two-thirds of the value of an animal which had died, while the society itself undertook to sell the carcass for its own profit.

A good many primitive systems of the same kind can be observed in France.

THE INSURANCE OF CROPS SOWN AND HARVEST PRODUCTS

We now turn from the insurance of cattle to the insurance of crops. In this case, there is a difference not only in the subject matter of the insurance, but in the nature of the economic problem which has to be resolved. The loss of cattle for reasons unrelated to epidemics is a normal economic phenomenon which can – in relation to large herds – be predicted with a high degree of accuracy and can be allowed for when preparing annual estimates for the cost of maintaining cattle. By contrast, bad harvests, losses due to hail damage or plagues of locusts, phylloxera or other pests are elemental disasters which are almost always unexpected and which do not occur by any means every year. Furthermore, damage to crops from these causes has always covered entire zones of land, extending over large regions, and has afflicted literally every farm in the region.

In all these respects, the financial and material losses relating to crops resemble losses of cattle through epidemics. Co-operatives that insure cattle are careful, as we have seen, to treat this as a separate problem, which in most cases they refuse to handle.

A certain degree of stability in the incidence of losses for risks of this kind may be achieved either by extending the areas of the territory insured to areas approximately equal to, or greater than, the territory of a state, or else by calculating the rates of loss by reference not to a single year but to a period stretching over several years. The rates of loss from damage caused by hail or by locusts or other pests will, if averaged out over the decades, give us comparatively stable coefficients and can provide a basis for insurance calculations.

Such are the special features which make the insurance of crops an exceedingly difficult problem not only for co-operatives but also for other kinds of insurance organization.

The most basic and the most important of these kinds of insurance, namely the insurance of crops against bad harvests, has scarcely been fully achieved in practice anywhere in the world. The only attempts we know of in this field were the storage of grain for

Co-operative Insurance

the possible 'bad year' which at one time took place in the Russian countryside and provided its food reserves; and the state insurance of rice crops in Japan. Among the schemes for crop insurance now being drawn up in the USSR, what merits the most attention is not the attempt to insure the crops of individual owners, but the measures aimed at enabling local co-operatives and local authorities to get resources to help with sowings, as well as the measures to combat the risks of famine.

We have considerable experience with regard to the insurance of crops against episodic mishaps due to damage from hail, destruction by frost, pests and all kinds of diseases. A significant proportion of this insurance work in the West has been undertaken by public bodies with legally enforceable powers as well as by private insurance companies. But besides that, a good deal of experience has now been built up in the field of co-operative insurance.

Summing up the experience of co-operative work accumulated in this field, we can sketch the following organizational outline of co-operative insurance against damage caused by hail.

Every member of the co-operative is required, during the spring, to declare for insurance purposes all the crops which he has sown which are to be insured by the association; and he must attach to his declaration at least a rough plan showing where these crops have been sown on his land.

The owner of the crops sown is himself allowed to estimate the anticipated yield of the harvest and its future value. The insurance premium is levied accordingly. The owner is warned, however, that compensation for any losses of crops due to hail damage will be paid at their real value and will not in any case exceed the anticipated value of the crops shown in his declaration. He is also warned that if only a part of the crops is damaged by hail, then the losses will be met up to an amount not exceeding the declared value of this particular part.

The owner must indicate in his declaration whether he is insuring grain by itself, or whether he is also insuring straw. Insurance is accepted for a period which usually runs from 1 May to 15 September, after which the harvest is presumed to have been gathered and the co-operative begins work on financial settlements with its insurance policy holders. Co-operative insurance is usually based on the principle of constant premiums. The level of premiums is determined on the one hand by reference to the climate in the area where the crops are sown; and on the other hand by reference to particular plants.

What was probably the earliest example of this kind of initiative where grain was concerned were the associations set up in France to combat the cockchafer (the 'Syndicats de hannetrage'), the first of which was formed in the Garron district of the department of Magenne. This syndicate, which was founded in 1866, served as a model for others. It set out to organize a concerted and vigorous campaign in all village settlements throughout its area, for the purpose of getting maximum results in alleviating the devastation caused by the beetle and its maggot. Landowners each paid 25 centimes per hectare into a common fund. The syndicate paid for all beetles collected in its area. It paid, at a price fixed by its council of management, out of money which had been raised, donated, bequeathed or provided in the form of grants. During the appropriate seasons it also sent out groups of workers, who were paid at fixed rates, to search for beetles in places which had been devastated by beetles. Lastly, it allowed farmers, peasant farmers and members of the syndicate to have the use, free of charge, of ploughshares, harrows, sprayers, and so on, to help them to extirpate caterpillars from the soil.

12
Associations Concerned with Land

Associations connected with the land hold a very special position among co-operative organizations. All the co-operative organizations that we have studied so far – for credit, purchasing, marketing, reprocessing, machinery and cattle-rearing – based their co-operative principles on the socialization of various parts of the economic turnover. By contrast, co-operation which is concerned with land does not affect either the productive, or any of the other, processes of the turnover of valuable resources within the household. Its purpose essentially amounts to the organization of the main territorial base for agricultural production.

The basic task of co-operation concerned with land is to organize the land area on which a farm will be set up; this applies equally to melioration associations and to associations for land tenure or for the joint purchasing or leasing of land. The most widespread of all these has been the melioration association.

The technical process on which it is based is essentially that of the so-called radical improvement of the soil. When we observe various pieces of land, we can notice that some of them are totally unfit for agricultural use; or that even if they are cultivated, they yield extremely negligible results. In most cases the reasons for the unsuitability of the land is that it is affected by an abnormal combination of the physical factors which determine fertility. Inadequate moisture will make it quite impossible for any of the ordinary cultivated plants to grow. Conversely, an excess of water leads to the swamping of the soil and to high acidity. Soil made up of quicksand which lacks humus to ensure its cohesion, or stony soil which makes cultivation difficult, or the formation of ravines which erode the surface of the land – all these abnormalities will hamper agricultural cultivation.

However, from the point of view of modern agronomy, none of these obstacles is insurmountable. Modern techniques enable us to dry out swamps, to irrigate deserts, to stabilize quicksands and to convert stone rocks into vineyards – in short, to turn 'empty places' into fertile fields and meadows.

From a technical point of view, melioration provides virtually unlimited opportunities for human genius. For an agronomist of the present day, the land is nothing more than a surface illuminated by the life-giving rays of the sun. But he is able to determine as he chooses the physical condition of this surface in the way required to absorb solar energy through the chlorophyl.

The crux of the question is: what will be the cost of this? And in our case, will the costs of melioration be repaid by the economic results of the improvements undertaken?

From the vantage point of an ordinary capitalist farm, the melioration of any piece of land is possible provided that the economic effect of this melioration – measured in terms of an increase of agricultural rent – is greater than, or at least equal to, the normal percentage return on the capital invested in the melioration. If the cost of the melioration is such that the rate of interest on the capital expended on it amounts to a sum greater than the increase in the net return which results from the increase in the harvest yield produced by the improvements made in the soil, then obviously the radical improvement which was envisaged cannot be profitable. The farmer will be obliged to abandon it, despite all its technical feasibility and attractiveness.

For this reason the state, which has an interest in melioration for the sake of the national economy as a whole, will quite often create special funds for the granting of credit on preferential terms, for the purpose of such projects for melioration. The reinforcing of ravines, the straightening out of river-beds, the fight against quicksands and other similar measures simply cannot be undertaken except on the basis of credit on preferential terms.

But economic calculations as to melioration are drawn up in a rather different way in the case of self-employed peasant family farms. In one of the early chapters of this book, we demonstrated through a detailed analysis that such family farms interpret profitability in their own particular way, which differs from the interpretation used by farms employing hired labour. We know that for a self-employed peasant family farm, net profit, from a book-keeping point of view, can be ascertained only tentatively; and that the earning of such a profit is not the actual purpose of such a farm.

A peasant family invests the labour of its members, and the capital that it owns, in its farm; and it seeks – while achieving the full reproduction of its capital – to derive the maximum remuneration for the labour which it has invested in the farm. The higher the remuneration per unit of this labour, the more successful the farm's organization should be deemed to have been. We should also remember that the main purpose of the peasant family farm is to provide for the family's annual consumption; and that therefore what matters above all is the remuneration for the year's labour as a whole and not the average remuneration per unit of labour.

Therefore, because its opportunities for profitably applying its labour are limited, the peasant household directs its effort to the growing of crops or to activities which, even when they yield a low remuneration per unit of labour, nevertheless enable the family's manpower to be more fully used, thus considerably increasing the remuneration of labour during the year as a whole. Accordingly, a peasant farm may expand its allotments by planting labour-intensive crops, even though they yield a lower remuneration per unit of labour, provided they make it possible to apply five times as much labour to the same piece of land. In just the same way, the peasant household will expand the area to which it can apply its labour, by taking land on lease at prices far above the ground-rent; and by buying pieces of land for sums of money which considerably exceed the capitalized rent.

These characteristics are also bound to affect the peasant household's capacity to undertake melioration. It can easily be understood that radical improvements, achieved through an expansion of the land area fit for exploitation, will of themselves help to satisfy the basic need of peasant households in regions which suffer from agrarian over-population – that is, the need for families engaged in economic management to extend the opportunities for applying their labour. Therefore, melioration work has an importance which is similar to the expansion of labour-intensive crops, the organization of earnings from cottage industries, and so on.

Our conclusions are especially important for the purpose of ascertaining the economic limiting factors of melioration. In the case of an enterprise run on capitalist principles, the rate of profit, corresponding to the usual rate of interest on capital, constitutes the economic limiting factor of melioration. Radical improvements will obviously make a loss if they produce an increase of rent below the usual rate of interest, calculated as a proportion of the capital spent on melioration; and such improvements will cease to have any point for a capitalist farm.

But for a peasant farm the economic limiting factors for melioration are considerably more flexible. A whole number of radical improvements, which would be beyond the reach of a capitalist farm, are possible for a peasant farm seeking ways of extending the application of its labour.

To put it another way, a peasant family farm is subject to a different limit, where the profitability of melioration is concerned. This limit cannot be conceived as some definite percentage return on the expenditure on melioration, since, as we have already demonstrated, the rate of return on capital in a peasant family farm is not a purely economic concept.

We are inclined to the view that, in general, this rate of return cannot be established by any purely objective calculations. It depends on the extent to which the families engaged in economic management are provided with the means of livelihood; on the amount of surplus work which they undertake; on the possibility or otherwise of finding alternative ways of expanding the application of their labour; and also on other conditions which are difficult or impossible to measure.

The only objective yardstick on which to base our approximation is, in our opinion, local land prices and, in particular, the prices for those economically significant tracts of land which melioration work itself serves to create.

The level of land prices paid by peasant households bears no direct relation to the level of net returns and is determined by precisely the same conditions as determine the opportunities for melioration.

Both the acquisition of land and the making of radical improvements to the land, provide increased scope for the application of labour by expanding the area of usable land. And it is obvious that a peasant household will not, for example, undertake the drying out of a swamped meadow if the cost of so doing is greater than the price for which meadowland can be bought in the district. On the other hand, if a peasant household seeking greater scope for the application of its labour buys fresh land at prices higher than the capitalized rent, then it is also obvious that any expansion of usable areas through the radical improvement of its own land will be profitable for the household, provided that the cost of this melioration work is below the selling price of the land, even though the increase of rent expected as a result is below the normal rate of interest on the capital expended.

It may be pointed out that most melioration work is undertaken with borrowed capital; and that the main condition laid down when credit for melioration work is granted is that the increase in rent must be used in order completely to pay off this credit. In cases where melioration work is undertaken, but where the increase in rent is not enough to pay even the interest on the borrowed capital, a self-employed peasant farm will have to repay the capital out of its 'wages'.

We cannot, of course, regard such a state of affairs as either favourable or desirable. But when we are considering the economic limits of melioration, we have to envisage that when such radical improvements are undertaken, self-employed peasant family farms will have to repay loans incurred for the purpose of melioration out of their 'wages', just as their 'wages' are used to pay rent for the land they lease; and just as 'wages' were used before the war to repay debts to the peasant bank for land which had been acquired at prices above the capitalized rent.

These, then, are the most general theoretical considerations relating to the economic opportunities for melioration in a peasant family farm.

The nature of such a farm does enable it to undertake melioration work of a kind which would be beyond the reach of a capitalist farm. Nevertheless, even in the case of a peasant family farm, the provision of cheaper credit for melioration makes it infinitely easier to carry out because such credit enables the farm to make these improvements without reducing the remuneration of the labour which it applies to the land on which the melioration work is being undertaken.

It follows from what has been said above that the first way in which co-operatives can assist melioration work by peasant households consists in the granting of cheap, long-term credit. The search for special credit funds which can be lent for long periods at low rates of interest is a direct obligation of the general co-operative credit system, which the latter has quite consciously set out to fulfil.

However, co-operative assistance to melioration work need not be confined to the financing of such work on preferential terms.

In most cases, melioration work will succeed only when it covers substantial areas of unusable land. The drying out and drainage of large areas of swamped meadows, the irrigation of substantial tracts of territory affected by drought – all this provides scope for melioration work on a large scale and thus significantly lowers the costs per unit of the area on which the work is carried out. This kind of expansion of the area of melioration work is very often essential not only from the economic, but from the technical point of view – since the achievement of results is technically possible only when the improvement is undertaken comprehensively and on a mass scale.

Because of this need for the comprehensive inclusion of large tracts of land covering several hundred or several thousand farms, co-operatives concerned with melioration work have certain distinctive features.

One of the main foundations of co-operation, as we know, is the principle that farms should join the co-operative voluntarily. However, this principle cannot be applied as a matter of course to a co-operative which is embarking on joint melioration work over a large tract of land. For a co-operative of this kind, it is essential for technical reasons that not a single farm on the land in question should be able to avoid taking part in the melioration work. Therefore, those in charge of melioration work have always striven to obtain legislative recognition of the right of melioration associations to have powers of legal enforcement when performing their public functions. The most important thing here is the association's right to insist on compulsory membership for the minority of farms on the land undergoing melioration, which are unwilling for some reason to join the association voluntarily.

Laws on melioration in various countries, accordingly, lay it down that if the majority of the farms on the tract of land intended for melioration form a special association for this purpose, then the dissenting minority can be compelled to undertake improvements to their soil and may not refuse to do the appropriate work on their land, or refuse to make the appropriate contribution to the cost of the melioration as a whole.

There are, it is true, various legislative provisions intended to safeguard the rights and economic interests of minorities; but only in so far as these minorities refrain from hindering the carrying out of the basic tasks. An owner is not exempted from sharing in the expenditure on melioration work unless he can prove that the melioration would involve him in loss. It must be noted that there are many foreign enactments on melioration associations which contain no reservations whatever with regard to compulsory participation. Nor does the relevant clause in our co-operative statutes contain any such reservations. It provides that:

> Where land melioration work, which has been entrusted to an association, cannot, for technical reasons, be carried out without affecting pieces of land whose users have expressed no desire to join the association which is being formed, or where the use of structures connected with the melioration work is bound to affect these pieces of land, then the persons and organizations which use these pieces of land can be required to join the association. For this purpose it is necessary that the members of the association should own at least two thirds of the land on which the melioration work is to be undertaken; and that the resolution which set up the association should have been approved by the votes of two thirds of all users of such land.

An association formed in this way needs, first of all, to obtain resources to carry out its work. Only in very rare cases do these resources come from the contributions of the members themselves; the passive funds of the association usually come from borrowed capital. Most commonly of all this capital, which is advanced on preferential terms, comes from the state or from some other public source. It is also not unusual for private capital to be attracted.

In order to attract private capital for the purposes of melioration, the most recent legislation on melioration lays down special preferential terms for the granting of loans for this purpose. More specifically, it gives them a right of priority: that is, it lays down that in the event of the non-payment of loans for melioration, claims for their recovery take precedence over claims for the recovery of mortgages, secured loans or loans of other kinds.

In some countries, the actual task of recovering annual redemption payments plus interest payments on loans for melioration is undertaken on the same principles as the collection of taxes. Sometimes the task is even performed by the same collectors, who bypass the association's accounts departments.

It is clear that such preferential terms for debt recovery are a very impressive guarantee that the debt will be repaid; and for every private owner of capital, the investment of resources in loans for melioration will be one of the soundest capital investments. This is bound to result in lower interest rates on credit granted for melioration.

In Western Europe, a loan advanced for the purpose of melioration is usually secured by the land of the farm which takes out the loan; and is therefore a loan in the nature of a mortgage. However, among those engaged in melioration work there is no unanimous agreement as to exactly which land constitutes the security for such loans. Some suppose that in the interests of easing the burden which loans for melioration impose on the peasant household, it is enough to grant such loans solely on the security of the particular land on which the melioration is being carried out. Others, however, regard this security as inadequate in precisely those cases where steps have to be taken to compel the repayment of the debt, that is, where melioration has failed. They insist on the need to mortgage not only the land on which melioration is being carried out but all the land belonging to the farm; although they recognize, of course, that such a demand may deter the peasants from undertaking the melioration work itself.

Under the land tenure regulations obtaining in the USSR, the mortgaging of debts in respect of melioration is impossible owing to the impossibility of selling land. Therefore, other kinds of security have to be found. Among these, the right of priority which we mentioned above does for the moment provide creditors with a certain guarantee.

When resources have been obtained and a detailed scheme for melioration work has been drawn up, the association embarks on its implementation, often with the help of its members' manpower, which appreciably reduces the cost of the work.

When the work has been completed, the association and its board of management do not cease to exist, since they are responsible for:

1. The winding-up of monetary relationships and the arrangements for repayment of the loans; and
2. Repairs, as well as the cost of making good any defects in the melioration which has been carried out (for example, the repair of overgrown ditches, damaged drains and so on).

If the melioration work involved irrigation, the association takes responsibility for managing the water supply for the irrigation network: it apportions the water supply and carries out work incidental to this, as well as raising the resources needed for the annual running costs.

Such, in the most general outline, are the basis of co-operatives for the purpose of melioration. Closely akin to them in their economic character are the associations set up for the purpose of jointly organizing land tenure on rational principles.

It quite often happens that the fields of various owners are dispersed and divided up into strips to such an extent that the sound management of a farm becomes extremely difficult. The area of 5 or 6 hectares which is used by the farm may be split up into numerous small plots which are often very far apart from one another.

Even in Western Europe the division of land into strips continues to be the scourge of agriculture. For example, an account of 19 ordinary peasant households in the neighbourhood of Weimar[1] revealed that in only one case was farming carried out on only three plots of land. All the remainder were based on five plots or more; while half the total possessed land consisting of ten or more plots. Because of this dispersal of land, the average distance between the fields and the farmstead in these cases exceeded 1.5–2 kilometres; and the distance to the most remote piece of land might be as great as 6–9 kilometres.

An even greater degree of land dispersal is encountered in the Russian communal system of farming, where it is not uncommon to find an allotment of 6 or 7 *desyatiny* [between 16.2 and 18.9 acres] fragmented into 40 or more strips whose form and location are of the most inconvenient kind from the economic point of view.

This situation with respect to economically important areas of land usually arises for historical reasons from one generation to another, as the result of the division and private transfer of land under a system of private land ownership; or else it arises as the result of communally held land being redistributed for the purpose of equalization. But it seriously erodes one of the most important of all the advantages of small-scale farming, namely the shortening of distances within the farm and the consequent reduction of transport costs within the farm.

The small-scale farm which remains small and organizationally weak begins to acquire all the disadvantages of large-scale agriculture in relation to transport within the farm.

The need to rectify these shortcomings in the organization of the land area often becomes so acute that it turns into a social calamity; and it leads to the intervention of the state authorities, who then arrange for fresh surveys to be made, for strips of land to be exchanged between owners, for boundaries to be adjusted, and so on. This leads to the drawing of new boundaries; and all the farms concentrate their land, so far as possible, into a single piece of land which can easily be run from a farmstead in the centre. The economic benefits of such an improvement in land organization are incalculable; and state policy in this respect is one of the most important forms of state assistance to agriculture.

Despite the fact that arrangements concerning land are usually made the responsibility of, and are usually undertaken by, legally empowered public authorities, this task is sometimes also within the capacity of co-operatives. At all events, we know of special associations in France and Germany (for example, the syndicates in Mourecourt and Renil in the department of Seine-et-Oise) which make it their business to secure boundary re-adjustments, to see that boundary marks remain intact and, most important of all, to build and maintain roads for agricultural purposes.

This latter work is fundamentally important because of the exceptional importance of transport within the farm. A rationally planned road network makes it possible, first, to economize on the land area used for roads, and, most important of all, it makes it possible to reduce the cost and labour involved in transporting agricultural loads.

Associations for the arrangement of land are identical in their economic nature with the melioration associations which we have just been studying. We may therefore suppose that it quite possible to apply to them all the same basic organizational principles which apply to co-operatives for melioration.

Both the types of co-operative which we have described have, from a technical point of view, transformed the land so as to facilitate its agricultural use. However, where land is concerned, co-operatives can also provide substantial assistance for the purpose of expanding land tenure. Associations for the joint purchase and leasing of land already have their own not inconsiderable history, and they have played an extremely beneficial part in organizing the national economies of agricultural countries.

What is of the greatest interest to us in this respect is the practice of collective leasing, which has become especially widespread in Italy. The idea behind it is extremely simple. The owners of estates who prefer to lease out their land instead of using it for economic purposes, will very often seek to avoid the burdensome and not very agreeable business of parcelling out their land among small peasant tenants. And they willingly hand over their land to special entrepreneurs who pay them a moderate rent and will then, in their turn, hand over the land in small plots to peasant farmers.

It goes without saying that entrepreneurs of this kind are ready, as it were, in their pursuit of high profits, to tear the shirts off the backs of the peasant farmers by fixing inordinately high rents for the particular pieces of land. In order to combat this kind of speculation at the expense of peasant farmers, societies for joint leasing were set up. These societies lease entire estates from their owners for long periods, for the purpose of subsequently distributing the land for use in accordance with co-operative principles. Cases have been known where agricultural workers jointly rent an estate on which they formerly worked.

Usually, when these joint leases are entered into, the tenants do not merely confine themselves to the signing of a general contract or to the distribution of the land between the participants in the co-operative. The fact of the joint lease will, of itself, have created the elements of a social nucleus, or even of a co-operative apparatus. The tenants will try to use this to the fullest possible extent: by instructing their board of management to organize joint purchasing and marketing, by setting up a butter-producing factory, by arranging for the joint use of machinery, and so forth.

A joint lease may quite often lead to attempts to set up a fully-fledged farming partnership, based on the labour of its members, that is, based on agricultural production. This is especially common when an entire estate is leased out, together with its buildings, stock and cattle and when the lease is granted to former workers on the estate.

However, the problem of joint agricultural production does not fit into the framework of the present chapter; and, since we believe this problem to be a matter of very great importance for co-operative life, we shall make it the subject of the next chapter, to which the reader is referred.

NOTE

1. Dr Herbst, 'Guts- und Betriebs Verhaeltnisse bauerlishen Gueter', *Thiel's Landwirtschaftlische Jahrbuch*, 1908, p. 381.

13
Collective Farms or 'Total Agricultural Co-operation'

The forms of agricultural co-operation that we have investigated all entail the more or less general collectivization of particular sectors of the peasant economy, thereby making them economically and technically stronger while enhancing the stability of the peasant family household.

We know that the peasant economy collectivizes precisely those sectors of its economic activity in which a large-scale form of production has significant advantages over small-scale forms; and it leaves to the individual family farm those of its sectors which are better organized in a small-scale enterprise. From this point of view, by no means all sectors of agriculture fall within the ambit of the co-operative system. Co-operative collectivism operates within very wide limits; but it is, nevertheless, subject to certain limits – which leaves a considerable area of activity to the family farm.

However, a good many co-operative theoreticians assume that all the forms of co-operative work which we have explained above are merely stages which will gradually lead to the complete socialization of all processes of agricultural production; and to the creation of large-scale collective enterprises, into which individual family households will be totally dissolved. This view was particularly widespread during the first years after the revolution, when agricultural collectives and communes constituted the pivot of our agricultural policy. Our collective farm movement has now existed for many years. We are therefore in a position to sum up a number of its results and ascertain, with a far greater degree of precision and clarity, its place in our co-operative movement.

As the reader will have learned from our first chapters, we are inclined to regard agricultural co-operation as one form of vertical concentration of agricultural production. From this point of view, the gradual detachment of particular activities in the peasant household's organizational plan, and the organization of these activities into large-scale enterprises, which, as it were, stand above the general mass of small-scale family farms, represents not a stage of development towards something else, but the affirmation of a principle in itself.

The promotion of full-scale collectives and communes would represent the replacement of the principle of vertical concentration by that of *horizontal concentration*. At first sight these two principles appear to be contradictory; and it would appear that since we shall have achieved horizontal concentration on a mass scale, therefore the principle of vertical concentration becomes superfluous. However, it is not difficult to show that this proposition is mistaken.

The reader may recall the theory, explained in Chapter 2, of differential optima, which maintained that for every type of co-operative work it is possible, on the basis of the economic and technical nature of the processes to be brought within the ambit of co-operatives, to determine the most profitable scale for the enterprises. From this, the reader will clearly see that the overwhelming majority of these co-operative enterprises will – owing to the scope of their work – have to operate on a territorial scale many times greater than that of even the largest communes or largest collective farms. Dairy farming associations operate within a radius of 3–5 versts [approximately 2-3.3 miles]. Those dealing with sugar beet and potatoes operate within a radius of about 10 versts [approximately 6.6 miles], while marketing, purchasing, credit and insurance associations operate over even greater distances, to say nothing of the alliances of agricultural associations. In other words, a collective farm, however large it may be, cannot replace the system of vertical concentration of agriculture but must – for the purpose of more fully achieving the concentration of agriculture – become a member of a local co-operative, just like a small peasant household. The only thing that an agricultural collective farm can replace is the small-scale machinery association. In this case, and probably in this case alone, the volume of a collective farm's economic activity does equal or exceed the optimal scale. But in other cases, the optimum scales are, as a rule, considerably higher than those of the very largest collectives.

Even if we were to suppose that literally all peasant households were ultimately merged into communes and were organized on optimum areas of 300–800 hectares [741–1,977 acres], this should in no way affect our basic system of co-operatives engaged in purchasing, credit, marketing and production, which would continue to be organized as before. The only difference would be that the membership of primary co-operatives, instead of being drawn from the small peasant households, would be drawn from communes.

A system of vertical concentration developed on a gigantic scale would, at its lower levels, be supplemented by the development of some measure of horizontal concentration. It must, however, be borne in mind that in order fully to implement the system of differential optima, agricultural communes would also have to detach a great many activities in their organizational plan and organize them in co-operatives on the scale of a larger agricultural enterprise.

In short, agricultural collectives can in no circumstances be treated as being the opposite of the system of agricultural co-operation. They should not replace, but should merely supplement, the system of primary co-operatives. Therefore, the question of collective farms comes down, in effect, to the question of who will be the members of the primary co-operatives: individual family farms, or large farms or collective farms. The choice would not be between *collectives* and *co-operatives*. The essence of the choice would be whether the membership of co-operatives is to be drawn from collectives or from peasant family households. And even as regards this question, the solution is by no means always or everywhere clear.

There is no doubt that with regard to buildings, stock, the use of animal traction and a great deal else, the scale of operations of a collective farm possesses advantages, which can be measured in substantial quantitative terms. At the same time, however, it cannot be denied that if one examines the very long list of economic factors which were once cited by David and other Marxist-Revisionists as factors defining the advantages of a small-scale farm, one finds some which are very significant from the organizational point of view, and where the advantages of a small-scale – or, more precisely, an independently owned – farm can also be measured in substantial quantitative terms. These include, in particular, those advantages resulting from the intensity of working effort, the effect of increased attention and the nature of managerial decision-making, etc.

Therefore, if one approaches this problem, not from the point of view of ideological aspirations, but simply by comparing the economic strengths of different social types of economy and their capacities for resistance and for survival, then the question ceases to be an issue of principle and it will resolve itself into a quantitative comparison of the ways in which the two kinds of factors mentioned above influence the overall economic result. And this result will, in all probability, differ in different regions and in different types of agricultural productive organization.

It must be assumed that collective farms can and will have a significant advantage over individual peasant households in regions that rely on extensive economic methods, where labour is organized in ways that are simple and easily amenable to mechanization, where most of the work is automatic and where enlargements of scale result in clear quantitative gains. Here, the only obstacles to the extension of collective farms will be social tradition and the resistance of the stronger peasant households, which do not want to lose their individual identities.

In any case, some features of collectivization, such as the shared use of tractors and, in general, the joint cultivation of the land, will undoubtedly become very widespread in the above cases.

By contrast, collective farms cannot become so widespread in regions that rely on intensive methods of cultivation – such as the cultivation of orchards and market gardens, potato-growing, dairy farming, poultry-breeding, and so on. In these cases, the mechanization and automation of labour would not produce large quantitative gains; while the quality of care and attention given can lead to a considerable increase in income. One exception is, of course, those units of land that were formed from the acquisition of orchards or market gardens already in existence. Such units, because of their established physical organization as large farms, are able to maintain cohesion and discipline among the collective farm's work force. In other words, the fate of collective farms formed through a fusion of previously independent peasant households will vary in different areas. And it must be assumed that they will, in the main, collectivize the joint field-work, or perhaps only the joint tillage of the fields and use of meadows, while leaving their individual members to run those sectors of farming which offer no high benefit from the enlargement of scale.

Such is our understanding of the nature of different types of collective farms and of their possible future. When we go on to study the organization of this kind of agricultural co-operation, we must first of all note the very great diversity of the most fundamental organizational types. The term 'collective farm' may refer, on the one hand, to full-scale communes where socialization sometimes extends to personal consumption or even to certain items of clothing. The term may, on the other hand, refer to co-operatives where the only joint activity of the members is the ploughing up of their land whilst everything else remains in individual user. In between these extreme types, there is a whole number of intermediate, transitional forms of organization.

In order to keep our attention on the main point, our analysis will concentrate on an average form of organization, that is, on the full-scale agricultural association or partnership, where individual consumption is maintained while all agriculture is socialized. In this kind of partnership, all the economic activities of the members are carried on within the framework of the partnership's economic enterprise; and the members themselves do not engage in any separate or individual enterprises of their own. This is a fully defined type of organization and its work will help us also to gain an insight into the remaining forms of collective rural organization.

Unfortunately, the very extensive literature on *collective farms* and other kinds of agricultural collective contains scarcely a single piece of research into their organization as agricultural enterprises. It is obvious that the organizers of the collective farm movement, while rightly regarding collective farms as large-scale agricultural enterprises, also supposed that they were totally indistinguishable in organization from ordinary large-scale agricultural enterprises based on the use of hired labour. They would therefore be amenable to the same organizational principles and the same structure of organization as those appropriate to state farms and to large-scale enterprises in general.

However, this is a mistaken view. The social nature of the collective farm, mainly with regard to the way its labour is organized and owing to the impossibility in principle of recruiting hired labour, requires a major departure from the type of structure which has been usual in large-scale enterprises. These differences can be reduced to three main characteristics. First of all, a collective farm, so far as its labour is concerned, is made up of a workforce of members of this collective farm. Since the recruitment of hired labour is ruled out in principle, this workforce determines the volume of economic activity for the collective as a whole. The nucleus of workers in a collective farm can develop the farm only within the limits of their ability to cope with their work during the peak periods of the gathering of the harvest and the ploughing of the land.

But this fact means that this nucleus of workers is condemned to a peculiar kind of unemployment at all other times of year, when the volume of work on various types of cultivation is considerably lower than during the peak periods. In order to counteract this, a collective farm has to make considerable changes in its organizational plan; and it must try to combine work on cultivation with work on the various sectors of the farm in such a way as to ensure that its manpower requirements over the year are spaced out as evenly as possible.

Secondly, if all the collective farm members are the owners of the enterprise as a whole with absolutely equal rights, it is exceedingly difficult to regulate questions of labour organization; and it is difficult to resolve questions relating to labour remuneration, the skills involved in particular types of specialized work, the allocation of duties and labour incentives. In order to prevent all the workers from being levelled down to the lowest common denominator, one needs either a spirit of enthusiasm throughout the collective, or else a system of labour organization and incentives that is capable of ensuring the necessary degree of hard work by all members of the collective.

A third and perhaps even greater difficulty centres on the authority of the head of the collective farm to manage or organize; and on the measures taken to uphold labour discipline in this kind of enterprise.

The dependence of an elected board of management and of its head on the people who elect them, and the impossibility of expelling members from the workforce as a disciplinary measure, considerably undermines managerial authority and thereby deprives it of the importance which it has, and must have, in any large-scale enterprise.

It is precisely these three organizational differences between a collective farm and all other large-scale agricultural enterprises which make it necessary for us to be especially careful when we explain how production is organized in a collective farm.

We must carefully scrutinize all the connecting links in the organizational plan as well as the structure of the economic organization of the commune and of all other types of collective. We have to analyse each aspect of the economic apparatus from the point of view of the three special features noted above; and we have to modify its structure so as to remove the organizational weaknesses of the collective farm which inevitably result from these special features. If we fail to take account of these facts, or fail to foresee their consequences, we shall inevitably get an incompetent economic organization which performs no efficient work and which fails to produce the economic results which could be produced by a large-scale enterprise based on hired labour and organized alongside it on the same scale.

The whole history of our collective farm movement from 1918 to the present points to the necessity of such an organizational analysis. The numerous failures of communes, the gradual and successive revisions of working instructions, the changes of organizational plan and, sometimes, the examples of malfunctioning, in the sense of the recruitment of a large number of hired workers – these are the lessons of the years which have gone by since the collective farm movement came into being. Very often, these organizational failings outweighed the advantages resulting from a large-scale form of production; and they led the collective towards a kind of economic organization far less effective than that of the individual peasant household.

Therefore, an examination of the special organizational features of a collective is, one might say, the fundamental question of the moment. It is the task of our research institutes to make a real and thorough study of this question. Work in this field has recently been carried out at our Research Institute of Agricultural Economics and some of its findings will be quoted here. We shall try to outline them as briefly as possible.

Under the conditions of a modern commodity economy, upon which a contemporary collective must necessarily base its programme, the main goal of the collective as an enterprise must be to produce commodities, that is, agricultural products for selling. This aim naturally requires the organized production of those products which, in the conditions of the particular farm and in the particular market situation, yield the greatest profit, that is, products whose production cost is significantly below the price which they will fetch on the local market.

The drawing-up of calculations and the organization of this market-oriented sector is the focal point of the entire agricultural enterprise.

However, no matter how profitable the market sectors of the farm may be, no agricultural enterprise can ever use the whole of its land area for these activities. The reasons are first, that in most cases, the agronomic prerequisites for production necessitate crop rotation. If all agriculture were reduced to the growing of one or two of the most profitable crops, this would impoverish the soil very quickly indeed. For the sake of stable fertility of the soil, crops have to be supplemented by crops of a different type, mainly intertilled crops and grasses. Furthermore, in the case of a whole number of goods for personal consumption and, to an even greater extent in the case of goods used for fodder, we constantly find a situation where their production costs on the farm are below their market prices. Therefore, as long as the organization of the output of the market-oriented part of the farm requires fodder for its cattle and food for its workforce, it proves far more profitable to produce and consume them within the farm. It follows from this that side by side with the commodity sectors of the farm we also have to organize two other sectors – those which produce for consumption and those which produce fodder. The establishment of a proper balance between these three sectors is the basic task when planning the organizational pattern of the farm.

As soon as this pattern has been decided on, and has been implemented through the allocation of land according to its economic purpose and through the establishment of the rotation of arable land, we can take further steps in organizing the structure of the farm. We have to calculate how many units of traction power are needed for the running of the farm. We have to compare our estimate of the farm's fodder production base with the needs of the animals used for traction and, after taking account of resources, we also have to organize them to provide fodder for the productive cattle. After organizing the fodder base, we have to organize the cattle-rearing; and having done this, we can go on to consider the provision of fertilizers and the organization of orchards and market gardens. Having thus planned in outline all the farm's main areas of productive activity, we can turn to the organization of equipment, ancillary production and buildings.

A large-scale capitalist farm, once it has made all the calculations listed above, will thereby have established practically all the details of its organizational plan; and having checked the plan's profitability by calculations made on the spot, it will finally decide how many workers it will hire during which months of the year. That is to say, only at the very last stages of its calculations, does it begin to consider how to organize its labour.

It can easily be understood that in our collective farms this system is turned upside down, since, in most cases, the amount of manpower is determined by the existing membership of the collective farm. A capitalist also takes account of his labour situation. But he does so only for the purpose of providing himself with the cheapest possible labour. And therefore, if he makes any alterations in his organizational plan when making his final calculations as to manpower, he does this only in order to carry out partial re-organizations of production so as to use manpower as far as possible during the times of year when wages are at their lowest. In this sense, large farms, particularly those relying on intensive methods – such as those growing beet, or potatoes or engaged in distilling, and so forth – were able to achieve an extremely high degree of organizational perfection. Large farms in Poland, Germany and Austria employed only a small number of permanent workers and based their organization of labour on the hiring, three times a year or sometimes only twice a year, of cheap peasant labour which was often brought in from a distance of several hundred miles. The peak of labour activity on the large farms was often consciously designed so as to be inversely related to the peak of labour effort in peasant households, i.e. designed to achieve the maximum effort at exactly the times when peasant manpower was idle and could be hired very cheaply.

As well as trying to choose the best times for this purpose, many large farms tried to achieve the same result by the choice of location. It was not a region with good soil which was considered most profitable for a large farm; but a region in which the farm's vast fields were surrounded by the most land-hungry peasant villages, with large reserves of surplus labour which they would have to offer very cheaply on the market, owing to their half-starved existence. One has only to examine the organization of labour at a whole series of sugar beet farms in the pre-war period in order to become convinced that this was so.

It can easily be understood that this organizational pattern is – both in principle and in practice – totally inapplicable to the organization of even the largest collective farm. The whole of the collective farm's manpower is, in principle, drawn from its permanent workers. The hiring of labour on a day-by-day basis is virtually ruled out; and therefore any unevenness of working effort points either to an inability to cope with the sowing of crops on the farm; or else it points to the involuntary unemployment of the workforce.

Therefore, after completing the calculations for the organizational plan, we must, at each stage of the plan, treat the question of labour organization as the main criterion; and when all the calculations for the organizational plan have been made, the plan has to be checked not so much by reference to the profits shown in the accounts, but by reference to the use made of the labour activity of the collective farm's members. Labour organization which, in a capitalist farm, was a factor derived from the organizational plan, is, in the present case, a fixed factor determining the organizational plan, which is itself drawn up mainly with reference to the way labour is organized. The organizer of a collective farm often has to abandon profitable methods of organizing the farm, simply because they are beyond the capacities of the main workforce of the collective itself. It often becomes necessary, purely for the sake of an even distribution of labour, to resort to the cultivation of less profitable crops and to rely heavily on the mechanization of labour in order to cope with the peak periods of sowing and reaping. It is precisely in collective farms that tractors and harvesters need to be most widely used, because they will be profitable even when their profitability on paper is negligible or even negative. A tractor, which cannot compete with the cheap labour of peasant ploughmen hired in the neighbourhood, nevertheless proves to be a powerful instrument for enabling members of a collective farm to deal with larger areas of arable land during peak periods and thus protect themselves against enforced unemployment during the remainder of the year.

Besides mechanization, there is a need for great diversity in the choice of consumer and fodder crops as well as cash crops, in order to expand the period of full activity. It is also not a bad thing to take advantage of the different varieties of plant and of the different periods within which different varieties of one and the same plant mature; and even to hasten or delay the sowing, in order to cope with periods when the work-load becomes severe. The theoretical literature on the organization of farms is at present merely raising these questions; and we have unfortunately not come across a single account of the organizational plan of a collective farm which consciously put this principle to the test. Nevertheless, practice is nearly always in advance of theory; and it provides us, in the most efficient collective farms, with examples of the actual implementation of such principles. We have to become aware of these principles and formulate them as definitive propositions concerning organization. It is to be hoped that, in the years immediately ahead, this research work will be carried out.

No matter how competent the calculations may have been with regard to the organizational plan of the collective farm as an enterprise, the plan can only be realized if the members of this collective work with at least the same degree of effort as is to be found in large-scale enterprises based on hired labour.

Work, as opposed to play, is described as work precisely because it is burdensome to the organism that performs it and requires a considerable effort and willpower if it is to be continued. If this effort is to be made, it necessarily requires some kind of incentive.

In a self-employed peasant family farm, the incentive to work stems from the needs of the family which have not yet been satisfied; and the degree to which they are satisfied depends on the degree of working effort. In a capitalist economy based on a piece-work system, the incentive to put effort into work is the wages, paid in proportion to the effort. In the case of work paid for by the day, which is very common in agriculture, the incentive comes from the coercive influence of the management, which has an interest in this working effort, and from the fear of losing one's job, or, in some cases, it comes from fines imposed for carelessness.

A major peculiarity of the system of incentives for hired labour is that where this labour is paid by the day, the incentives often serve not so much to encourage the performance of the work for which the wages are paid, but rather to encourage the appearance of effort in this work. On the other hand, the incentive for the labour of a peasant household consists in achieving the results of the work; from which it follows that the peasant's work will, even with an identical amount of effort, be more purposive and/or more productive.

Some ideologists of collectivized agriculture argue that the advantages of economic collectivism are due precisely to its superior system of incentives for human labour, in comparison with those of a capitalist economy. They suppose that a collective, because it is based on the work of its members, will thereby acquire the same exceptionally powerful incentives which exist in the self-employed family farm; and that this will be further reinforced by the psychology of 'working in a collective'.

We shall not deny that in certain cases this claim may prove valid: in small collectives which are spiritually closely-knit or are inspired by some social or religious ideal, the incentives to work may be stronger than in any other kinds of economic organization. It is impossible, however, to deduce any general rules from these particular cases. In those numerous collectives where spiritual ties between the members are slight and where there is no strong enthusiasm for common action, the system of incentives just noted will weaken; and the principle will begin to prevail which can be crudely expressed by the statement: 'Why should I work harder than my neighbour when he and I get paid the same?'

When products are divided up equally among the mouths to be fed and when enthusiasm is lacking, the work of collectives differs little from work paid for by the day. And since the collective's coercive will is always less energetic than the will of a one-man proprietor striving for maximum profits, cases may arise when the system of incentives in a collective is considerably inferior to that of a capitalist economy, based on hired labour paid for by the day.

The mind and will of a collective are always less active and more sluggish; they hardly ever provide scope for the sort of intuition which is so important in any kind of entrepreneurial work. The will of the owner of a capitalist enterprise and of the head of a self-employed peasant family ensure the cohesion of the organizational plan and its steadfast implementation in practice. But the collective will is weaker, first of all, in the organizational and entrepreneurial sense, and secondly, in the coercive sense: since the agents of this will – personified by the board of management and by others who have been elected – are too heavily dependent on their voters to possess steadfast resolve.

Apart from these weaknesses of collective organization with regard to production, it is not an easy matter to ensure even the mere cohabitation of families within a collective farm. Hence the not infrequent disintegration of agricultural partnerships, which are quite viable from the economic point of view, but are torn apart by internal dissensions.

We must, of course, recognize that enthusiasm based on an idea or on religion can sometimes prove more powerful than all the shortcomings just enumerated. But one can hardly design a collective agricultural movement which is intended to be long-lasting and on a mass scale, by relying on enthusiasm.

For these reasons, the organizers of collective agriculture need, first of all, to resolve two fundamental questions: how to establish labour discipline within a partnership; and how to create within the collective a psychological incentive to work harder. The solution to these fundamental problems holds the key to everything: all the rest is, in effect, only a technical problem.

When we scrutinize the organization of particular activities in the sphere of collective agriculture, we see numerous attempts to mitigate the shortcomings of the system pointed out above. First of all, those sectors which are least successful when run by a collective or large-scale farm are detached from the enterprise as a whole and are left to be run individually by the families who have joined the collective. Thus, only a very few collectives at the present time socialize the housekeeping or prepare food on a communal basis. Cases are not uncommon where socialization extends only to field-crop cultivation, the care of the meadows and forests and the grazing of cattle. Cattle-rearing as such, as well as market gardening, often remain under the control of the individual family household. One can even find examples in literature where the term 'agricultural collective' referred to nothing more than the joint cultivation and harvesting of the land held on lease, while the cultivation of allotments continued in all other sectors of the farm on an entirely autonomous and individual basis.

There is no doubt that all the exceptions just listed did a good deal to strengthen the stability of collective partnerships as an economic instrument. But they nevertheless required further special measures in order to eliminate the shortcomings, mentioned above, of collective organization.

Even where the collective nature of agriculture amounts to nothing more than the common cultivation of the fields and harvesting of the crops, the organizers of a partnership will have to think about how to strengthen the managerial will and about how to reinforce the incentive to work.

Collectives usually choose a particular individual or collective board of management, endowed with a measure of absolute authority in their executive work, in drawing up the plan of work and allocating workers for each day's duties. In short, they perform a role corresponding to that of the proprietor in a capitalist farm. Members of the collective farm are, however, often subject to an extra discipline backed up by a system of fines or deductions from the products distributed. However, these sanctions may remain merely theoretical if the authority created in this way lacks any adequate incentive to maintain the strict standards which would be demanded by the proprietor of a large enterprise concerned about his personal profits. In collective farms, this personal incentive has to be replaced by some kind of system to provide those incentives. Its absence may prove to be disastrous.

Thus, in the collectives which are most communistic in spirit, all the products obtained are distributed in kind among the families, according to the number of mouths to be fed. A family with one worker and three members who are not able-bodied will consequently receive twice as much as a family comprising two workers and no other members of the household. It can easily be understood that such a system of income distribution is the one least of all conducive to the encouragement of work for the sake of personal profit.

A slightly more effective system is that of distributing products among the workers only. In their desire to reinforce the personal interest of an individual in the success of his work, the organizers of such collectives have proposed a variety of arrangements for distributing the income obtained.

The most effective arrangement of all from this point of view has been the one whereby an enterprise run by a partnership is managed, in a formal sense, according to the capitalist model or, more precisely, on the substantive model of the Rochdale system of co-operation. This means that every member of the collective is regarded as a worker and is paid wages according to the amount of work which he has actually done, sometimes on a piece-work calculation. (It is true that he is not always paid in cash: a considerable part of the wages is credited to his account.) All produce obtained from the fields is deemed under this arrangement to belong to the entire collective, which sells the produce wholesale. The part of the produce which is handed over in kind to members of the collective is paid for by them either in cash, at market prices, or by deductions from the wages due to them. The result of this type of system is that at the end of the year, the collective will usually find itself with a considerable amount of profit at its general disposal. Some of this profit is paid into the collective's social funds, for the renewal and expansion of the collective farm's capital, and for purposes relating to the common benefit as well as for cultural and educational purposes. Another part of the profit is distributed among the members according to the amount of work contributed by each of them to the collective farm.

The system of distribution just described is sometimes provided for in the statutes of the collective farm; but in some cases, it is decided annually by a resolution of the general meeting.

The system just examined does provide an adequate system of work incentives. At all events, it does so to at least to the same extent as can be seen in a capitalist farm. However, with this system of financial settlements, a collective ceases to have the idealistic character of a free commune.

What matters more to us is not the psychology or ideology of the members of a collective farm but the economic realities of its existence and the kinds of collective which really are able to exist in the conditions of a commodity economy and competition without being propped up and without needing any shield to protect them. For this reason, the principles described above need to be put into practice and have in fact been put into practice by those collective farms which have achieved actual stability. However, these methods of organizing production incentives bring good results only if the members of the collective have themselves been well chosen. It may confidently be said that half the success of a collective farm enterprise depends on its personnel. It has long since been pointed out in literature that there are four basic requirements for the cohesion of the work force in a collective.

The members of the workforce must:

1. Have an adequate material interest in collective agriculture; that is, they must derive from it an income at least as large as what a member was getting before he joined the collective.
2. Be accustomed to agricultural work;
3. Be of more or less the same educational and social level; and
4. Possess an adequate social and technical capacity for collective economic management.

The size of the collective, and the area of land that it cultivates, must be sufficient for it to benefit from the advantages of a large-scale economy and a complex division of labour. At the same time, they must not be so large as to require a complicated system of management.

The size of collectives established in practice by the Dukhobor sect was of about 40 peasant households. In Italy, the size of collectives was equivalent in practice to an average of about 150 people; that is, about 60 peasant households. Our own usual figures are slightly higher. But if the size is large, this gradually gives rise to the problem of higher transport costs within the farm as well as the difficulty of maintaining a unified management.

The size of a collective's land tenure is defined by the size of the collective. The piece of land is either acquired, or formed by other means.

The impossibility of arbitrarily expanding or reducing the area of collective land tenure and the simultaneous need to make the fullest possible use of the manpower of its members will determine the size of the collective; and the collective will therefore be compelled – in contrast to all other types of co-operative – to reject the principle of the unrestricted admission of new members. For if the area of land tenure remains constant, a new member, if he is not to become a superfluous burden, will be able to join the collective only in order to replace a former member.

The impossibility of altering the area of land tenure likewise explains why collectives allow the hiring of workers from outside, in the event of members falling ill or leaving, because the remaining workforce is physically unable to cope with the whole land area. Moreover, as we have already noted, such hiring is sometimes allowed in collectives for the additional purpose of helping their members at the time of the harvest and at other times when manpower organization is crucially important, if the organizational plan is unable to ensure that work is evenly spaced out over time.

One difficulty which it takes great effort to overcome is that of creating a managerial will to run the collective farm as an efficient economic enterprise. For the purpose of running the farm's affairs, a board of management is chosen. It is sometimes supplemented by a special technical committee whose members take charge of particular sectors of the farm as a whole. It has a leader/chairman who is, as it were, a dictator in matters relating to work, embodying the authority of the collective.

The technical committee is made responsible for, among other things, the planning of the crop rotation and the planning of other aspects related to the organization of the farm run as a partnership. This organizational plan must, however, be confirmed by the general meeting of members of the collective. As soon as it is agreed, it is implemented by the collective's managerial bodies.

As we have already noted, work in these collectives is organized in the same way as in capitalist farms. The collective's chairman, acting on the instructions of the technical committee, compiles daily work directives which are binding on members of the collective. Members are paid ordinary wages for this work. Everything produced by the collective is deemed to belong to the collective as such. Part of it is sold to members at market prices; another part is sold outside, after the payment of wages and running costs. The income earned by the collective provides it with a common income. Part of this is used for the repayment of debts, amortization and the formation of capital; while the other part is distributed between members, according to the amount of work performed by each of them.

In addition to this, the collective's managerial bodies perform all the tasks which arise in other types of co-operative; and from this point of view they also provide services for the individual peasant households of members, if any of these remain.

In a formal sense, these are the tasks and rights of the collective's management. But the issue is not what is put on paper or decided by a show of hands at a general meeting. What is important is not the plan or the form of organization, but how the plan is implemented. The organizer of a collective farm has to overcome an endless number of centrifugal forces and frictions which arise in the functioning of this type of co-operation. An exceptional personal authority is needed in order to avoid all manner of complaints of unfairness and favouritism – if only over the allocation of heavy and light work between members. Many collective farms have gone to ruin owing to the fact that no one wanted to perform heavy unskilled work when other members were doing lighter or more agreeable work.

This difficulty has, it is true, been somewhat mitigated by the use of different scales in the remuneration of labour. The misfortune, however, is that if these scales are based on the rates laid down by the state or on the wage rates paid in practice, the result is to make heavy manual labour even more disagreeable, because it is least well paid.

Only the personal authority of the person elected to manage the work, and his influence on his partners, can guarantee smooth work where this is concerned. However, the very fact that personal authority is so important can itself give rise to a good many other dangers, because this personal authority can easily turn into personal dictatorship; and the collective can gradually be converted into the personal enterprise of its leader. Vacillations between these kinds of Scylla and Charybdis are indeed the basic problem which has given cause for disquiet in the development of collective farms.

We shall not dwell, in this outline, on the question of the organization of capital in our collective farms. But it must be noted that from the organizational point of view, this is one of the most positive features of the collective farm, since it is precisely the large-scale types of farming which enable us to exploit the land with only half or less than half of the capital required for peasant households based on the minutest parcels of land. As early as 1913, before the Revolution, A. Minin put forward the idea of collectivization as the only solution to the problems of those peasant strata who possess land but lack the means of production and money capital. These problems still remain at the present time; and they lead us to believe that collectivization and collective farms are a valuable form of organization for agricultural production in precisely those regions where land is relatively abundant but where there is a clearly visible shortage of the means of production.

To this we might add that, in those cases where the collective's managerial will is weak and where labour incentives are also relatively weak, collective farms will stand to lose least in those regions where the forms of production are simple and mechanical and where the opportunity to make widespread use of tractors and agricultural machinery will have a disciplining effect on the workforce concerned. If these conditions hold good, we can easily understand that – in the case of collectives which are formed not through the exploitation of the farms of former landed estates, but which were created and developed through a combination of peasant households – the widest opportunities exist in the grain-producing regions, which rely on extensive farming methods, in the south and south-west of our country and in Siberia. This view is fully borne out by the cases of success of large-scale peasant collectives in these regions.

However, if we are to make them more stable and raise the productivity of their work to the maximum, then we must, in all seriousness and with a full sense of responsibility, place a further question on the agenda. This is the question, not just of studying the organization of collective farms, but of considering how best to draw up their organizational plans. This must be done not merely by applying the rules and guidelines of capitalist agriculture, but by developing the kinds of autonomous creativity which stem from the organizational foundations and principles of collective farms.

14
The Basic Principles of Organization of Agricultural Co-operatives

An attentive reader of this book who has read through the preceding chapters might justifiably criticize us on the grounds that while we have often described in great detail the way the co-operative system operates, we have nevertheless said practically nothing, or very little, about the societal apparatus that conducts these operations.

This is a real omission; but it was quite deliberate and the reasons are as follows. We wanted to describe, in the clearest way possible, the particular kinds of co-operative work and their connections with various aspects of the peasant economy. Therefore, we had to adopt a somewhat abstract approach, i.e. for the sake of clear explanation, we divided up co-operative work into a number of sections which were apparently independent of one another. In actual fact, by no means all the kinds of co-operation listed above exist in separate forms. In the overwhelming majority of cases, a variety of co-operative functions are performed by one and the same working apparatus. A Siberian butter manufacturing partnership undertakes the work of a consumer co-operative and the selling of oats; while a credit association will often undertake the buying, selling and insurance of cattle. In other words, the actual social mechanism of co-operatives is determined not so much by its technical functions, but rather by the general conditions of development of the co-operative movement in the country or region under review.

We must never forget that we are dealing with a co-operative movement of the peasantry – that is, with a broad social movement, which is constantly developing and moving from one phase into another, which exists in differing legal and economic conditions and which creates its organizational forms in accordance with these conditions and with the state of its own development.

Particular kinds of co-operative work are merely particular manifestations of what is essentially a single social movement. Therefore, in order to ascertain what kind of social apparatus the peasantry relies on to undertake the co-operative functions which it needs, how this apparatus is organized and what are the motivating forces behind it, we have to set about analysing the peasant co-operative movement in its totality. We have to think carefully about the forms in which it has developed, and to see what kinds of organization it is able to create at different stages of its evolution. This kind of comprehensive examination of agricultural co-operation is especially necessary and important in the context of our Soviet economy, where a centralized co-operative apparatus, which forms part of a planned economic system, must necessarily be examined as a single whole.

The first thing that strikes us when we study this subject is the extraordinarily chaotic way in which forms of co-operation developed historically; and the paucity of any conscious design in resolving the problems which faced co-operative organizers in everyday life.

One great historian, Vasilii Klyuchevskii, when describing the epoch of reforms under Peter I, tells us that the reformer did not, in effect, have either a plan or a general programme for state action. His specific reforms arose out of particular needs; and were in most cases in the nature of technical improvements made necessary by military or other considerations of the particular moment. Nevertheless, as he moved with tireless energy from one particular problem to another, Peter unintentionally achieved a system whose coherence should be ascribed not so much to his subjective awareness or creative will, but rather to the objective needs of the country's economic development and of the development of the state.

It is highly likely that all thoroughgoing social and economic reforms have precisely this attribute of a spontaneous and irresistible social current. At all events, this was certainly true of the development of capitalism, which arose without any organizational blueprint, without any inventors and without the master plan of any social architect.

It may at first seem that the development of the co-operative movement was a direct exception to this rule. It is no accident that all co-operative calendars abound in portraits of Robert Owen and Fourier and that the name of Raifeizen is held in unfailing respect. It might be easy on this basis, therefore, to argue that co-operative ideas were conceived by the brilliant minds of social reformers long before the moment when the first co-operative appeared on earth; and that a social system which had been quite consciously conceived and worked out *a priori* became, over the course of time, translated into reality.

And indeed, if one disregards such everyday forms of co-operation as partnerships or the joint tilling of land, one has to admit that the first steps in the co-operative movement, at least here in Russia, were due to the energetic advocacy of major figures of the 1870s, who spread an awareness in the countryside of ready-made co-operative systems; and who implemented them with the steadfastness and consistency of an enlightened absolutism.

However, these initial efforts in themselves were not accompanied by the necessary objective preconditions for their implementation; and they therefore failed for a long time to bring any genuine co-operative movement to life. But when, owing to the development of a commodity economy in our countryside, the necessary preconditions were fulfilled and when co-operatives had gained a practical mastery of the experience of co-operatives in the West and had learned from the first pioneer enthusiasts, the situation began sharply to change.

The practice of co-operation – as it spontaneously developed in breadth and depth and came to include more and more areas of work – began to come up against organizational problems of a kind which required immediate solutions but which had not been envisaged by any of the existing co-operative theories. The practical need to solve the given problem was so great, and the problem itself was usually so sharply brought into relief by a particular situation, that the solution was sought, without any subtle philosophizing, by ordinary co-operative practitioners who possessed no outstanding capacity for abstract thought. They did nevertheless have a practical grasp of the matter as well as an organizational hunch.

There then came a period when co-operative theory came to follow in the wake of co-operative practice and developed, not in an *a priori* manner, as in the past, but in the manner of an *a posteriori* theory.

It was precisely in this way that the entire unified structure of co-operatives in our country, as well as the whole of the co-operative marketing system and most of the system of co-operative production, were created. The general outlines of this edifice have for the most part already been completed; and the need has naturally arisen to unify its individual parts and to select the best organizational forms from among the mass of forms which arose in a semi-spontaneous way. It is therefore only now that we are beginning to become aware of the nature of the co-operative organization and to formulate the theoretical foundations underlying the forms which have evolved in practice.

It can easily be understood that in the light of this *a posteriori* analysis, we find a good many regularities which, like Columbus's egg, would not have been hard to foresee earlier on.

First of all, we have already noted more than once that from the point of view of its work as a middleman, the structure of the co-operative apparatus with regard to technical organization and with regard to the relationships between primary co-operatives and the various levels of associations, largely corresponds to the structure of the commercial apparatus which co-operatives were intended to replace.

Earlier on, when discussing the organization of marketing co-operatives, we showed how the co-operative movement, as it proceeds step by step to capture the market for a particular commodity, will replace the cattle-dealer by a local co-operative, will replace the local trader by a co-operative alliance of an intermediate kind and will replace an export bureau by an all-union centre. At the same time, the practice of co-operatives is, of course, guided not by the wish to imitate its adversaries, but by economic necessity. So far as the commodity circulation is concerned, co-operatives are confronted with the same national economic problems as those which confront commercial capital. It is, therefore, natural that they should solve these problems in ways which are identical, since they are objectively the most effective.

The age-old experience of commercial capital led it to subdivide the process of commodity circulation into a number of primary processes; and, in relation to each of these processes, or category of closely interconnected processes, to create its own special apparatus of the appropriate size and capacity. These same primary processes in the commodity circulation continue to operate after the market has been brought within the co-operative system; and therefore, in the majority of cases, objective considerations of profitability require that the processes should continue to be subdivided into the same groups, with a special apparatus for each of them, similar to those used by commercial capital but based of course on co-operative principles.

The special features of co-operative work will often make it possible, and sometimes make it necessary, to change the structure of such a working apparatus, either by expansion or curtailment. But in their general outlines, owing to the identical national economic conditions by which they are governed, the organizational patterns of the co-operative and the commercial apparatus remain very similar. But it is a different matter with regard to the development of co-operative structures which were not created to replace a previously existing capitalist apparatus but were created from scratch, thus bringing new and previously unknown economic processes into the life of the countryside. This group of co-operative undertakings includes credit co-operatives which, even though they do not mean the first appearance in the countryside of credit as such, do represent the introduction into the countryside for the first time of organized credit. This group also includes co-operatives for cattle insurance, societies of stockbreeders, machinery users' associations, melioration associations, and so on.

However, the same overriding economic principles prevail. The work of co-operatives is subdivided into a number of categories of operations which form part of a technical whole; and for each such category a working apparatus is chosen of the kind which can carry out these operations cheaply and efficiently.

Thus, for example, all the essential operations of a co-operative credit system naturally break down into three groups; and they are respectively performed by three kinds of organization:

1. The verification of the solvency of peasant homesteads, the granting of loans, the supervision of the way they are spent and the recovery of loans, as well as the business of persuading the peasant population to deposit its money with co-operatives – all these operations require an apparatus which works in the closest possible proximity to the peasant, which constantly monitors his economic activity and which can sensitively adapt to it. This type of work can, of course, only be carried out by a small-scale district co-operative.
2. However, the apparatus of a small-scale local co-operative does not have the capacity for successful financial management of the kind undertaken by banks. The maintenance of stable credit balances requires a much greater volume of credit turnover and a more highly skilled staff than small-scale co-operatives can afford. Therefore this kind of operation – as well as operations for the granting of credit to co-operatives which do not themselves grant credit and which work in marketing, reprocessing, and so on – necessitates the organization of a co-operative apparatus with a wider area of activity and a large economic turnover. This constitutes the basis for the formation of co-operative associations of the second, i.e. provincial [*gubernskii*], level, to whom the operations just mentioned are assigned. These associations are usually made responsible for providing local co-operatives with the services of a specialized staff for the purpose of introducing, and giving guidance on, co-operative methods. They do so through the setting up of a special institute of instructors.
3. But the apparatuses of provincial associations – no matter how

large their turnovers may be – will never carry sufficient weight on the international capital market; and in order to consolidate their influence in the latter area, co-operatives have to set up special central apparatuses for the purpose of establishing links with the world money market and also in order to achieve the necessary guidance of co-operative financial affairs on a national scale.

The working apparatus of the co-operative credit system can therefore be subdivided into three components: the local co-operative, the co-operative association and the national co-operative centre. The distribution of work between them is determined on the basis of a more detailed analysis of the nature of each co-operative operation.

When we subdivide the process of the commodity circulation on the market into their individual components, we can – by the same methods as we used in an earlier chapter to determine the optimal areas of activity for trading apparatuses and the optimal siting of equipment for reprocessing – make use of this organizational analysis in order to elaborate a very detailed model of a co-operative apparatus functioning at two, three or four levels, which can easily undertake the organization of co-operative marketing or purchasing.

The same kind of logically elaborated models can easily be designed for all sectors of co-operative work; and it is possible, on paper, to create the most detailed design for the co-operative apparatus in the USSR as a whole, consisting of tens of thousands of co-operatives, associations and centres, each of them specialized and ideally suited in theory for the work they perform.

However, the logical elaboration of an organizational idea is not the same thing as its implementation. The crux of the matter concerns the methods of realizing it, and not the methods of its logical elaboration. One relevant example, which can also cast light on practically all the basic questions of co-operative development, is the history of our marketing co-operatives, whose real-life existence began only between 1913 and 1915.

The idea of the co-operative marketing of goods produced by peasant labour could not, of course, be described as a new idea. Already from the eighties of the last century, Russian social thinkers had recognized that it was necessary and desirable to bring the marketing operations of the peasant economy into a co-operative system. But only in the most recent times has this logically simple idea been implemented. For there was something which hindered the realization of an apparently simple theoretical notion. A mere belief in an idea or even the fervent preaching of the idea were not enough to get the idea implemented in practice – until concrete organizational methods were found for implementing the idea within the existing economic and social environment.

The art of politics is, first and foremost, the art of implementation. Even the most exalted social ideas and even the most ambitious plans, no matter how carefully and thoroughly worked out, possess a real value equal to zero from the point of view of economic policy, in the absence of an appropriate social environment and of methods of implementation. The organizer's most important art is to correlate the goals which have been set with the available forces and resources; and it is this which is most often forgotten by various people who draw up ambitious projects. The mere recognition of the need for co-operative marketing, or even the drawing up of a schematic plan for its organization, will not by themselves give birth to co-operative marketing; although any co-operative movement can, indeed, tackle the problem of marketing, once it has achieved the necessary maturity and organizational strength.

When we turn over the pages of old reports, we can see that the pioneers of co-operative marketing – for example, in relation to flax – gave first priority to goals which even the powerful co-operative organizations of today recognize as being beyond their capacity and which they regard as nothing more than a very remote ideal.

Members of Russian co-operatives, having acquired a certain organizational experience from working in credit, consumer and butter manufacturing co-operatives, approached the organization of marketing on co-operative lines by seeking to transpose their organizational skills to this field. It was assumed that it would be possible to start work by organizing small-scale local co-operative units, which would unite the peasants for the purpose of the joint reprocessing and marketing of products of good quality. The work of these primary units, so it was thought, could make organizers familiar with the organizational and technical aspects of marketing, give them an opportunity to gain the necessary experience and inculcate an awareness of co-operatives among the masses. It was further assumed that these primary units could, when they had grown stronger, combine into associations, first at the district level, and later in the form of an All-Russian Association, provided that marketing co-operatives were able to win the necessary mass support and achieve the necessary economic strength.

A local flax-processing partnership, an association for the collection of eggs, a group of bee-keepers or poultry-breeders or an association for the marketing of corn – these were to be the first steps in the development of marketing co-operatives, as envisaged by members of co-operative societies in the 1900s. Such was the *logical* way of forming marketing co-operatives. But *historically*, they developed along a different path – by organizing co-operative marketing not through specialist co-operatives but through a general system of agricultural credit co-operatives, built from above, not from below.

In order to grasp the reason why history diverged from 'logic' in the development of these co-operatives, one has only to make a thorough investigation of the goals of the flax co-operatives, of the economic conditions in which they were set up and of the organizational resources which they had at their disposal.

The economic goal of the organizers of flax co-operatives in the years 1913–15 was to drive commercial middlemen out of the market and to ensure that flax fibre passed directly from the producer to the consumer. Having eliminated commercial capital from the market, co-operatives must themselves undertake the national economic functions, which this capital performs: that is, they must collect from among the peasants the flax fibre which is scattered in small quantities among individual households, they must gather this fibre into large stocks, sort it according to quality into various grades and send it to spinning mills for processing.

If, from the national point of view, co-operatives perform this work better than commercial capital, then co-operatives will capture the market as well as the considerable middlemen's profit which accrues to the commercial apparatus and which is handed back to the peasantry once the market is organized on co-operative lines.

It must be noted that this capture of the market has to come about not by virtue of any privileges or government directives placing the flax trade under the monopolistic control of the co-operative system but by virtue of the intrinsic superiority of the co-operative apparatus over that of commercial capital. Real and lasting victory can be achieved only through organizational superiority and the better performance of one and the same economic task.

It is hardly necessary for us to show that the co-operative system has all the formal prerequisites for winning such a victory. As an association of raw material producers, the co-operative system can curtail the profits of middlemen. It is thus able, when selling the goods which it has assembled, to reduce wholesale prices almost down to the level of prices in the bazaar – thereby wiping out any competition by commercial capital. By taking a crop from the peasant when it is still virtually growing on the root, the co-operative system can guarantee the absence of adulteration as well as proper sorting of a kind which is flexible and responsive to market requirements. As a result, a co-operative commodity will necessarily enjoy a reputation for good quality.

Such are the theoretical advantages which can enable the co-operative system to win the victory which it seeks. It should be recognized, however, that the economic and social environment in our countryside considerably undermines these advantages. The mass of peasants who are not very cultured and are far from being aware of their own interests will often find themselves dependent on local traders; they will be exceedingly resistant to co-operative propaganda; and they will take their flax to a co-operative only when, in so doing, they see an immediate material advantage by comparison with a sale in the bazaar.

But it is by no means always possible to offer the peasant, from the very first year, a price which is appreciably higher than that of the bazaar. The reason is that, owing to the small turnovers of co-operative primary units and the high overhead costs incurred by inexperienced organizers, owing to the refusal to cheat over weights or to lower standards when grading the flax or allow its adulteration, the cost of co-operative flax is higher than the cost of the flax assembled by the commercial apparatus. This cost can be covered only if the market recognizes the intrinsic quality of co-operative flax and values it more highly than commercial flax of the same grade.

It is, however, impossible to expect this higher valuation so long as particular consignments of sub-standard co-operative flax appear on the market. A good and lasting reputation can be built up only as a result of the appearance on the market of co-operative goods on a mass scale, which are distinguished by having constant grades and a constant superiority. This is possible only when there exists a powerful and highly developed co-operative organization which has an impressive quantity of the commodity at its disposal. There arises therefore a kind of vicious circle. A co-operative system can develop only when it offers undoubted advantages to the peasantry. But it can offer such advantages only when it has developed and become sufficiently strong.

We already had occasion, in the chapter which dealt with questions of the organization of co-operative marketing, to make a detailed analysis of the questions which we are now expounding and of ideas for resolving them. What we are now saying is to a large extent a repetition of what we said there. However, this set of ideas is absolutely crucial to us at this point – because we have encountered this vicious circle twice in the course of our argument; and it is extremely important for our entire movement that the vicious circle should be broken and that it should, from an organizational point of view, be eliminated.

The only way out of this vicious circle is for co-operative marketing to be developed not through the creation of new, small-scale units which are inevitably doomed to perish, but through placing the organization of this matter in the hands of powerful co-operative economic apparatuses which already exist. Co-operative exports can be successfully organized only through the entry into the market of special, national organizations which have a large quantity of a product at their disposal and which represent a substantial factor on the international market. Unless this condition is satisfied, a co-operative organization, which as a rule has a limited grasp of commercial technique, will get lost in the markets; and the market will fail to appreciate the inherent advantages of co-operatives, namely the absence of adulteration and the careful grading.

But the lack of awareness among the masses involved in the co-operative system makes it necessary for co-operative marketing to provide the population with an immediate tangible benefit from the very first year. A commodity of high quality costs more to produce; and when offered small consignments, the market will not respond to this quality by offering higher prices. Consequently, it is impossible to pay the peasant a high price for what he has produced.

It might seem attractive to build up, let us say, a co-operative export organization, by adopting a gradualist approach – beginning with the creation of special co-operatives at the first level and afterwards, as they develop and grow stronger, combining them into local associations and finally setting up an all-union centre. But for the reasons just given, this approach has to be rejected. Co-operative centres have to be set up simultaneously at the start of the operation; although local work can initially, pending the setting up of specialist associations, be assigned to general co-operative territorial associations.

By making use of already existing multi-purpose local agricultural associations, we can immediately start setting up the vast apparatus which will be in a position to tackle the truly enormous national economic task of achieving the mass transfer of a product from the producer directly on to the world market. The fact that our co-operative marketing system was created in precisely this way in the years 1913–15 and 1923–24 indicates that this is the only correct path.

However, when a large number of previously independent organizations are joined together for working purposes, we have to remember that the success of an export operation is possible only when the activities of these organizations are made subject to a single plan and a single organizing authority.

In commerce, the will of the individual entrepreneur who acts promptly, often on the basis of intuition, has an overriding advantage over the collective will of co-operative organizations, which, as well as competing in the market, are also obliged to justify every step that they take before their fellow members. This has the effect of weakening the co-operative system; and the effect itself needs to be mitigated through the granting of special powers to central co-operative bodies.

At the same time, this managerial and executive authority – with regard to the fixing of prices, the technical conditions in which grading is carried out, the packing and dispatching of goods and the making of agreements and financial settlements – must be exercised in such a way that local associations can, after giving full executive power to the centre, still exercise a controlling influence on the running of the associations' affairs.

From a logical point of view, we envisage these centres as organizations with a narrow specialization, created separately for each association. However, from a chronological point of view, the functions of such a managerial centre, when work is just beginning, can and will be temporarily undertaken by a general co-operative centre which is already in existence.

Thus, for example, when the marketing of flax, eggs, hemp, etc. was being organized on co-operative lines, the functions of a national association were at first performed, in 1913–15 by the Moscow People's Bank [*Narodnyi Bank*], and in 1921–22 by the Rural Co-operative Alliance [*Sel'skosoyuz*]. Only when the work had got into its stride were the separate specialist centres set up to meet the practical requirements. Moreover, this process of splitting off proved successful only when the time for it was ripe, that is, when the volume of work in the particular specialist sector had increased to a sufficiently high level to finance the maintenance of a special apparatus; and when cadres with adequate authority had become available for this purpose.

It can confidently be said that the question as to when the situation is right for splitting off part of a general multi-purpose organization is always complex; and is one of the most difficult questions to resolve from the point of view of co-operative tactics. It is quite obvious that by the time that this splitting-off occurs, the work of the corresponding department of the integrated centre must involve a commodity turnover on a sufficiently large scale to make it possible, given the usual commercial charges on turnovers, to cover the estimated costs of maintaining the trading apparatus. It is also essential that a base should have been created, both at the level of the co-operative association and at lower levels, which can guarantee supplies of the appropriate quantity and quality of goods; and that a commercial clientele should exist so as to provide an assured and stable market for the goods. Only this system of social and economic links, and especially of co-operative links, can provide a basis for the future work of the specialized centre. There must be the same care when testing the financial basis for the operation. Lastly, the development of this work requires the formation of a strong, knowledgeable group of leaders, as well as a nucleus of office personnel.

The business of taking all these circumstances into account and ascertaining whether they are in fact operative represents the exceedingly difficult task of developing the co-operative system in a harmonious fashion. The co-ordination of the rate of development of each of the elements, and the work of ensuring that they develop in a dynamic as well as a harmonious way, is the most difficult problem which confronts the leadership of the co-operative movement.

The co-ordination of the expansion of co-operative turnover, of the organizational development of the co-operative apparatus and of the expansion of the financial base of co-operative work – these are the necessary guarantees of success in a co-operative operation, or indeed in any kind of large-scale economic operation. Any disharmony in the parallel development of these elements will inevitably have serious economic consequences. It is precisely for this reason that the growth in the volume of operations or growth in the financial base which underpins them will, at a certain stage of expansion, inevitably require that the apparatus which handles these operations should be converted from a department of an integrated association into a specialized centre. Any attempt to force this process or, conversely, any attempt to hold it back will unavoidably lead to negative economic consequences.

The agricultural co-operative system in the USSR was in fact built up in accordance with the theoretical principles just set out. It is, moreover, characteristic that this process occurred twice over in almost identical forms, since the system of agricultural co-operatives which developed during the period from 1913 to 1920 was then wound up and made into a branch of the national consumer co-operative system. A new co-operative system was developed entirely afresh during the period from 1922 to 1926.

Between 1915 and 1917 the following organizations were detached, one after the other, from the commodity department of the Moscow People's Bank: the Central Association of Flax Growers [*Tsentral'noye Tovarishchestvo L'novodov*], the Potato Growers' Association [*Soyuzkartofel'*], the Fruit and Vegetable Growers' Association [*Plodovoshch'*], the Hemp Association [*Pen'kosoyuz*], the Egg Co-operative [*Koyayitso*] and the Grain Co-operative [*Kozerno*]. Besides that, a Co-operative Insurance Alliance [*Koopstrakhsoyuz*] was set up; as well as a butter manufacturing co-operative organization [*maslodel'naya kooperatsiya*], which had no centre of its own, but operated through regional associations.

Beginning in 1922, during the Soviet period, the following were detached from what was originally the integrated Rural Co-operative Association controlled by the state [*Sel'skosoyuz*]: the Flax Growers' Centre [*L'notsentr*] (which also included hemp), the Potato Growers' Association, the Fruit and Wine Growers' Association [*Plodovinsoyuz*], the Butter Producers' Centre [*Maslotsentr*], the Poultry-breeding Association (*Ptitsevodsoyuz*], the Tobacco Growers' Association [*Tabaksoyuz*] and the Grain Producers' Association [*Khlebosoyuz*]. The meat producer' centre, as well as the centres which deal with sugar-beet, cotton and bee-keeping are close to being split off. The old Rural Co-operative Alliance is more and more being turned into a purchasing organization of agricultural co-operatives (for machinery, seeds and manure) which combines these functions with those of a centre, pending the creation of separate organizations. The Insurance Alliance and the banking centre, in the form of the All-Union Co-operative Bank [*Vsekobank*] which has been joined to the Transit Co-operative Bank in Riga and the Moscow *Narodnyi Bank Limited* in London, complete this system.

The result as of 1926 is that we have the following system of co-operative centres which rely on a common network of local multi-purpose associations as well as on their own special systems (see Figure 13).

No one has ever denied that it is necessary, and may even perhaps be technically inevitable, for separate specialist centres to be set up at a certain stage in the development of the co-operative movement. The idea of specialization was recognized very early on. But this question assumed a rather different significance with regard to local associations and primary co-operatives, since it arose in a totally different context. Specialization by primary co-operatives in the countryside provoked, and to some extent continues to provoke, very fierce argument.

There are some co-operative activists who think that because of the shortage of co-operative activists in the countryside, and because of the lack of resources and the poverty of the peasant population, we cannot set up specialized co-operative organizations in the countryside along the lines of capitalist enterprises or along the lines of co-operatives in Western Europe. Indeed, we saw in the quite recent past how some local co-operative organizers thought it desirable to do away with the separate existence of consumer and craft co-operatives and to merge them with agricultural co-operatives. They also thought it possible and desirable simply to create uniform general associations in the villages, which would

The Basic Principles of Organization of Agricultural Co-operatives 237

Figure 13

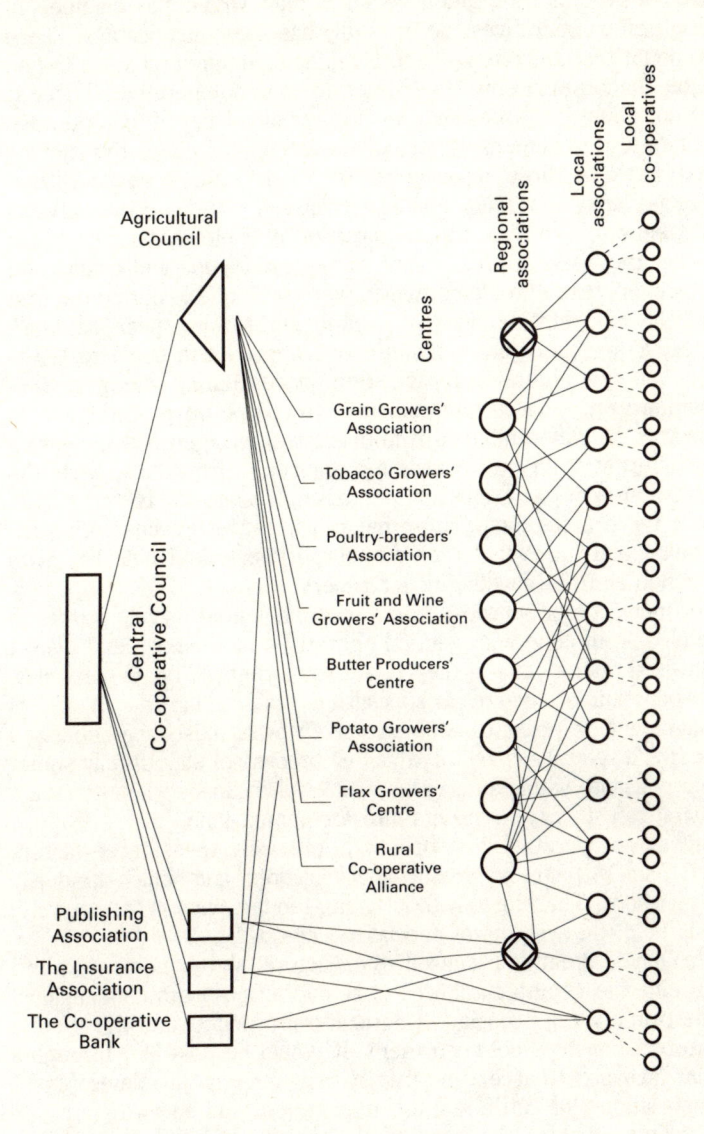

Note: This and the following diagrams have been taken from the report made by G. Kaminskii, Moscow, 1926.

undertake all forms of rural co-operative work.

The desire for integration, even if only within the confines of agricultural co-operatives, undoubtedly has some justification. There is no doubt that it is often extremely difficult in our rural areas to find suitable leaders and employees even for one co-operative. There is also no doubt that because of the small scale of co-operative operations, it is immensely difficult to set up and cover the running costs of two or three separate organizations within a single village. There is, lastly, no doubt that specialist associations do not always find it easy to agree on the demarcation of their various functions; and that they have a great many causes for friction and conflict.

However, these problems, which were insuperable during the first years of co-operative activity, will begin to disappear as work proceeds; and there are a number of factors which begin to tell in favour of specialization. In the first place, during the period of transition from reliance on middlemen to co-operative production, it transpires that different kinds of co-operatives engaged in reprocessing have different optimum radiuses for collecting their goods; while the optimum area of one credit and purchasing association has to include two or three areas for potato-grinding and butter manufacturing co-operation and an even larger number of associations dealing with inspection and the handling of machinery.

Secondly, a multi-purpose board of management usually proves to be well able to cope with general operations of a simple kind; whilst specialized operations – if they relate to marketing, or, in particular, to production – require a specialized staff within the board of management, in order to deal exclusively with these operations.

Lastly, a specialized sector which embraces not all, but only some, of the peasants within an area, is usually disinclined, when its affairs are well run, to pay its profits into a common fund.

But as co-operation develops, it is precisely these three factors which usually begin to favour specialization; and which gradually bring it about. Therefore, without going too far, we still have to rely, in this field, on systems of specialized co-operatives.

We have already discussed the factors and conditions which determine the geographical area of activity of co-operatives engaged in the primary reprocessing of agricultural products and the system of interrelationships between them. It would be possible, through a similar analysis, to ascertain what system governs the development of associations for cattle-rearing, melioration, and so on.

It follows that at the base of the co-operative system, at the lowest level, we will have a number of small-scale, specialized co-operatives for reprocessing. These will vary with regard to their optimum radiuses for collecting their goods. We will also have a considerable number of inspection, machinery users' and animal-rearing associations, as well as collective farms, etc. It is highly probable that for the purposes of finance and credit, all of these – just like the households included in the co-operative system – will belong to a purchasing, marketing and credit association at district level. This association will itself be sited close to a village with a bazaar, which is the commercial focus of gravitation for the whole area. The optimum area for this association must correspond to the area of natural gravitation for the local bazaar, which constitutes a primary cell of the economic system. Thus, the question of the specialization of the primary unit is, within the limits shown above, theoretically predetermined.

There is, however, no doubt that the way in which primary units develop, as well as the eventual pattern of the local co-operative network, can and will vary in different areas, depending on the local economic situation. It is possible that in our country, as in the Latin countries, the connecting link between all local primary co-operatives will in the course of time be provided not by a credit association but by a non-commercial alliance, something like an agricultural society.

A much more difficult and complex question is that of the stability and specialization of local associations. Let us take, for example, a provincial city on which in former times private capital focused its commercial and credit operations as well as its technical operations for grading and secondary reprocessing. As a rule these operations were all concentrated in one centre for each area. In other words, there was in this particular case no economic necessity for different radiuses for different types of work. Hence the extreme stability of local multi-purpose associations. So great was this stability that in relation to the potato, fruit and vegetable, tobacco, sugar beet and other co-operative systems, there were and are cases where, despite the existence of a specialist centre, the association continued on a multi-purpose basis.

In this case, the only possible argument in favour of specialization is that when the operation has sufficiently expanded, and when the splitting of the association into two involves no increase in overhead costs, there is a certain benefit to be gained from specialization within the managerial board and from a strengthening of ties specifically with the specialized primary unit. The greatest incentive for the separation of specialized operations is undoubtedly the desire of specialist primary co-operatives to keep for themselves the entire profit from their operations and not to pool it in a common fund. However neither of these incentives is always enough to detach a sector from the unified branch to which it belongs. This usually happens only when operations have assumed a wide scope.

An even more complex question is that of associations of the provincial [*gubernskii*] and regional [*oblastnoi*] types. In most cases – since they occupy an intermediate position in between local associations and centres – they have no economic *raison d'être*, since an analysis of the market apparatus demonstrates that, as a rule, only three tiers of marketing apparatus are economically necessary. At the same time, questions of organization and, most important of all, questions of representation and of links with state organs in the province [*guberniya*], make it absolutely imperative for an influential co-operative apparatus with full authority to be based in the provincial town, so that it can, within the framework of our planned economy, maintain contact with the Provincial Executive Committee and with the provincial planning agencies.

The need for this is very keenly felt; and it often makes it necessary to create associations not in the district [*uyezd*], but – despite a certain loss of contact with the primary units – in the province [*guberniya*]. From an economic and organizational point of view, however, this method of solving the problem does not always lead to good results. Life has not yet clearly pointed to any other solutions. Representation may, possibly, be delegated to the district associations within a provincial town. It is also possible to set up non-commercial organs in the provinces (such as inter-co-operative councils, councils of agricultural co-operative congresses, and so on) to perform these functions relating to representation and control. Nor can one rule out the possibility – which makes the best sense from the economic point of view – that credit operations will be transferred from the district associations to those of the provinces. The main consideration should be that of practical experience.

The most important work relating to the organization of exports and imports, as well as work of an inter-regional nature, has remained in the hands of the centre. These centres represent types of system which are by no means identical and which are determined by the differing natures of the economic processes which they organize. Some of them, especially in those cases where the work of the primary unit involves reprocessing, are not only specialized themselves, but rely on other specialized sectors. Cases in point are the Butter Producers' Centre, the Potato Growers' Association and to some extent, the Fruit and Wine Growers' Association. Figure 14 gives an idea of their structure.

Most of the other associations do not have specialized local branches. They rely on the common network of multi-purpose associations, which use their apparatuses for a whole range of different operations, and place them in specialist hands only at the highest level, where operations are assigned to a specialist centre.

The Basic Principles of Organization of Agricultural Co-operatives 241

Figure 14

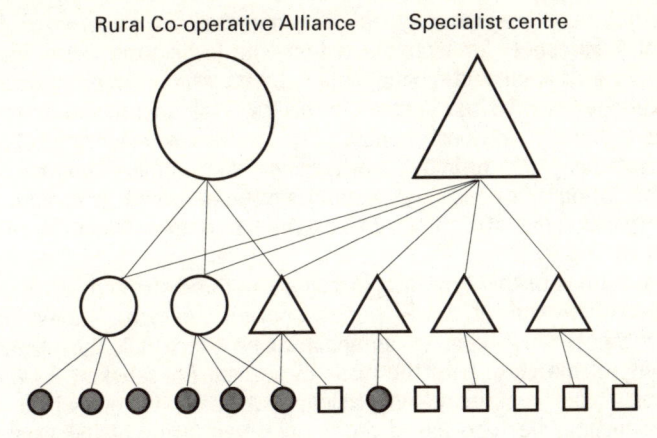

Rural Co-operative Alliance   Specialist centre

○ Multi-purpose associations
△ Specialist associations
● Local multi-purpose agricultural associations
□ Local specialist associations

Figure 15

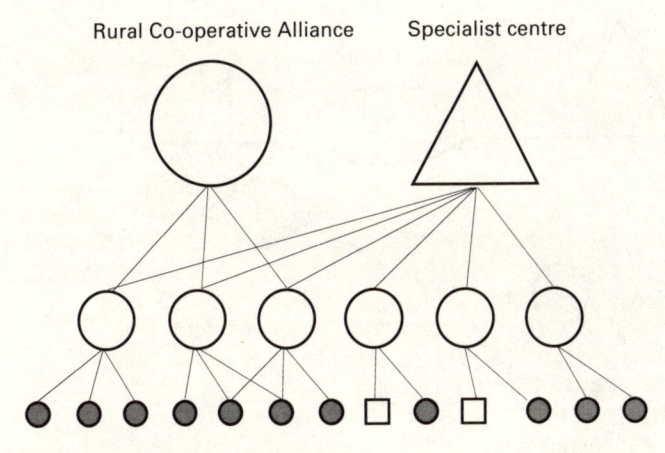

Rural Co-operative Alliance   Specialist centre

○ Multi-purpose associations
● Local multi-purpose agricultural associations
□ Local specialist associations

In their case, the organizational pattern assumes an entirely different form, as shown in Figure 15.

Lastly, there are some types of co-operative activity, as in the case of sugar beet, for example, where the trade turnover is itself always of a local character and cannot, by its very nature, involve a physical operation by the centre. In this case, all operational activity remains entirely in local hands. The centre retains control of organizational and training work, and also retains control of representation and concludes general agreements with contractors. The organizational structure of this type of co-operative system is shown by Figure 16.

Associations of this type graphically show the distinctive organizational characteristic of the Soviet co-operative system, namely the role of co-operative centres as apparatuses which link the planned economy of the state with the mass of peasant households. As time goes on, it becomes increasingly clear that in the system which we are examining, i.e. one based on co-operative forms of the vertical concentration of agriculture, it is the marketing co-operative systems which wield the greatest organizational power. By uniting the peasantry for the purpose of marketing the commodities which are of the greatest importance for each of them, these co-operatives affect

Figure 16

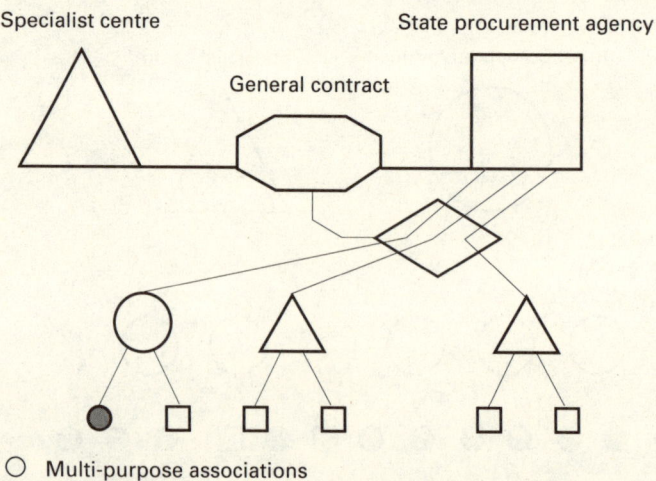

○ Multi-purpose associations
△ Specialist associations
● Local multi-purpose agricultural associations
□ Local specialist associations

the most important and most vital parts of the household. The ultimate task of the marketing co-operative system is to protect the peasant's money earnings, that is, figuratively speaking, his wages. It is here that he is most sensitive, it is here that his interests are focused. And it is our very strong belief that in the context of agricultural co-operation, it is only the marketing co-operative systems which need to be built entirely on co-operative lines at all levels.

Centres that deal with purchasing, insurance, transport, technical matters, publishing, electrification and even credit can be built up without regard to the forms and attributes of the co-operative movement. They have no need for meetings of local representatives or councils or collegial boards of management. They can simply act as technical offices, established on a shareholding company basis; and in all probability the co-operative system will only gain from this. The nerve centres of the co-operative movement, the social forces which constitute its component parts, run through the marketing systems. And it is precisely these marketing systems which must also serve as a link with the state agencies responsible for the planning of agriculture.

One can immediately realize that this is so by comparing the kinds of discussion which take place, on the one hand, at meetings of representatives of the Insurance Societies or the All-Union Co-operative Bank [*Vsekobank*]; and, on the other hand, at a representatives' meeting of, let us say, the Flax Growers' Centre or the Butter Producers' Centre.

In the overwhelming majority of cases, the marketing systems act in the USSR as export organizations which operate on international markets and which – despite the advantage which they derive from our state monopoly of foreign trade – are obliged to wage a hard struggle against the capitalist giants. This fact, as well as the general importance of the marketing systems, obliges us to make a detailed examination of these systems.

We already know that the experience of the Siberian butter manufacturing co-operatives, of the Flax Growers' Centre and of other large-scale co-operative associations gives us every ground for supposing that no significant success in the organization of co-operative exports can be achieved without the entry into the market of specialized nation-wide organizations, which have large quantities of the given product at their disposal and which represent a substantial force on the international market.

It would not be effective to adopt a gradualist approach when setting up these co-operative centres. The centres must be set up all at the same time, from the start of the operation; and pending the setting up of specialist associations, local work may be assigned to general co-operative associations of the territorial type. However, despite the apparently obvious character of these propositions, such centralized management can only be achieved gradually and with great difficulty.

It is recognized that one of the foundations of the co-operative movement is the spontaneous initiative of the population. The local co-operative unit is the primary source of co-operative life. It is here that new plans come into being. It is here that co-operative life is created. And it is also here, so it would seem, that the will of the population which has spontaneously joined the co-operatives is expressed. Local co-operatives are compelled by technical necessity to combine into associations. But these associations only exist in so far as co-operatives exist at the grass-roots and act in accordance with their will. In order, however, to ensure the success of the operations which they undertake, co-operative associations very often have to interfere in the work of local co-operatives, placing them within the constraints of a plan of action and making them subject to directives laid down by the association's management.

The same thing occurs in relations between the central association and the local ones. Thus, for example, central co-operative associations for the marketing of agricultural products make precise rules for local units with regard to the grading and packaging of the product; and they assume the entire responsibility for entering into transactions, and for fixing prices and payments. Local agencies are left with the sole task of implementing these plans.

This state of affairs would seem to deprive the local organization of any power of its own and to turn it into a subsidiary of the central association. However, there is no alternative to this, because the transfer of managerial power to the local association and the conversion of the centre into an agency of the local alliances – into a kind of office which handles commissions – will deprive the co-operative system of the strength and active competitive power which are essential to it.

We are, therefore, faced with an almost insoluble problem: how to uphold the power of the population which has spontaneously joined the co-operatives, while at the same time establishing the power of the central associations, whose unity and independence are alone capable of ensuring the 'commercial' success of the enterprise. It is a complex and painful question, especially when opinions are divided; and the solution will vary for different types of co-operation and in the differing conditions of time and place.

We already noted in one of the early chapters of this book that the principle of the direct responsibility of the organs of a co-operative organization to the members which it serves is a fundamental principle of the co-operative movement. In the absence of this principle, co-operation effectively ceases to be co-operation.

One extremely interesting example from this point of view is that of the old Central Association [*tovarishchestvo*] of Flax Growers, created in 1915 by four major local associations, which already had considerable experience of co-operative work, including work for the marketing of flax. During its first year of existence it was in a weaker state than its individual members – with regard to its moral authority and, still more, its financial position.

Hence the original rules of management, adopted at the founders' meeting. These rules had provided that the Central Association could undertake sales only subject to the gradings and valuations laid down by the local associations. Some local members even tried to establish the principle that every transaction entered into by the Central Association must be ratified by all local ones.

However, already by the end of the first year of operations, the disadvantages of this system had become obvious. Furthermore, the centre had, after a year of successful work, gained the necessary experience and authority in the eyes of local associations. Indeed, the Central Association's expansion and the recruitment of scores of new members significantly reduced the relative influence of each individual member. In short, the Central Association became the symbol of the entire flax co-operative system; and it acquired real strength, considerably greater than the strength of each local association. Following the second year's work, the rules on internal relations were revised yet again; and the fundamentally important principle was laid down of the organizational unity of the flax co-operative system. This principle was also the working foundation of the new Flax Growers' Centre [*L'notsentr*] created in 1922.

Just as the Central Association of Consumer Societies [*Tsentrosoyuz*], the All-Union Co-operative Bank [*Vsekobank*] and certain other co-operative organizations constitute federations of independent local associations which have relations, which are freely entered into on each occasion, with the centre which they have created, so in the same way the Flax Growers' Centre constitutes a single organization, merged with local associations. It is engaged in the marketing of flax and subject to a single management, which has freedom to take decisions with regard to the processing, grading, pricing, sale and transportation of the commodity.

This idea was particularly clearly expressed in the 'Rules for the co-operative marketing of fibre', which were confirmed by the Council of the Flax Growers' Centre on 31 October 1924.

Under section 8: The associations are to undertake seasonal operations for the co-operative marketing of fibres in accordance with the present rules on the basis of applications and plans which they have presented to the Board of management of the Flax Growers' Centre and which the latter has confirmed.

Under section 17: The maximum valuation of fibres is laid down by the Flax Growers' Centre. It is communicated to the associations when procurement begins and it may be altered by the Flax Growers' Centre at any time. Throughout the entire time when the fibre is being collected, the associations must make regular weekly reports to the Flax Growers' Centre as to what its acceptance prices actually are; and it must immediately inform the Flax Growers' Centre by telegraph of any rise or fall in prices. If the Flax Growers' Centre decides that for commercial reasons it cannot agree to the prices reported to it, it then has the right, after a certain period, to discontinue the further collection of the fibre, having at the same time informed the association of the prices which the Flax Growers' Centre can guarantee. Any association is, however, entitled to accept at its own risk the amount of money by which the acceptance prices exceed the guaranteed prices. It may, therefore, after informing the Flax Growers' Centre, continue its collection – but with the proviso that the amounts of money paid out by the Flax Growers' Centre are to be fixed by reference to the value of the fibre at the guaranteed prices. Consignments of fibre which have been procured on these conditions may not – for a period of 50 days after the Flax Growers' Centre has paid for them – be offered for sale by the Flax Growers' Centre at prices which fail to cover the association's actual expenses together with the deductions due to the Flax Growers' Centre.

Under section 20: Fibre which has been accepted from a producer by an association, which operates according to the rules of co-operative marketing, is deemed to have been irrevocably transferred at the moment when it is handed over by a co-operative organization.

Under section 21: From the moment when information is conveyed to the Flax Growers' Centre concerning fibre which has been accepted by a co-operative organization, the Flax Growers' Centre becomes liable for any loss of the fibre through natural disaster or accident. The association continues to store the fibre and is liable for losses caused by carelessness or improper handling during storage.

Under section 22: The association undertakes to store the fibre in its own warehouses until it receives an order from the Board of Management of the Flax Growers' Centre. The association dispatches the goods as directed.

Under section 23: The right to dispose of fibre accepted by associations for marketing is, under the present rules, vested solely in the Flax Growers' Centre. It exercises this monopoly for the purpose of marketing the fibre in the way most advantageous to the producer. For these purposes, the Flax Growers' Centre has the right in its own name to sell the fibre, to mortgage it, to give receipts for it, to move it from one place to another, to have it reprocessed, and so forth.

Under section 24: When delivered by the producer, the fibre is valued at its local market price and, on being accepted, is provisionally paid for at a price no higher than this valuation. The final payment for the fibre is made when it has been sold by Flax Growers' Centre in the manner indicated below.

Under section 25: Final payment for the fibre is made at the end of the season. It covers all consignments, including consignments of non-co-operative goods which have been sold by the Flax Growers' Centre up to 1 July (sections 5 and 7). These are treated as a single quantity of goods.

Without in any way encroaching on the autonomy of local associations and co-operatives in other areas of their work, the centralization of work for the collection of flax fibre by means of these rules has been successful in meeting the need for a unified apparatus and for freedom of manoeuvre in managerial decision-making.

The principles evolved in the flax trade have also been adopted by other branches of co-operative marketing.

It is obvious that their implementation can be described as 'co-operative' only if two conditions are satisfied. First, the centralized management examined above should be undertaken by an elective body, which has the complete trust of the mass of co-operators and whose actions fully reflect their wishes. Secondly, in cases where the wide-ranging powers of this body extend to executive functions, then its goals, as well as the planning of its activity should be approved by a body closely related to the co-operative rank and file. In other words, the greater the powers vested in the boards of management of associations and centres, which represent the characteristics of the co-operative system as an enterprise, the greater must be the role of councils and meetings of representatives, which embody co-operation as a social movement. This raises a crucial problem: the problem of combining the commercial flexibility of our organization with its co-operative character.

The basic organizational principles for the building of agricultural co-operation, which have just been examined, can therefore be set out as the following basic propositions:

1. The organizational forms of agricultural co-operation are basically determined by the economic and technical nature of the national-economic process or processes which are to be organized on co-operative lines. The vertical concentration which co-operation seeks to achieve requires the creation of a whole system of commercial and industrial organizations which perform differing functions and which are built to optimal scales which vary for each different function. In their overall system, these organizations are fairly similar to the capitalist organizations which have historically evolved in the same area of operations.
2. None of these organizations should have any economic purpose of their own. They consist of bodies set up by the peasant farms for their own benefit. They must be managed in accordance with the basic co-operative principle, i.e. that the managerial bodies of every organization are directly responsible to the members whom they serve. In order to enforce this responsibility, special bodies must be set up in the form of general meetings, meetings of representatives, councils and auditing commissions which, taken together, define the aims and methods of work.
3. It follows from the two principles just explained that it is logically possible to devise at the very beginning a theoretically ideal system of co-operative organizations, specializing in every kind of co-operative activity and mutually linked within a system. However, in an actual historical context, the immediate fulfilment of a logically elaborated plan will prove impossible. The various forms of co-operation have to be built up with a degree of gradualism, commensurate with the historically growing power of co-operatives as a social movement. It is, as a rule, necessary to begin new types of work by utilizing the resources and organizations which already exist. Only gradually, after the scale of operations has expanded, can this work be handed over to a specialized apparatus, should it prove technically necessary. At the same time, however, there are some marketing operations which have to be conducted on a large scale from the very beginning.
4. When particular sectors of co-operative work are detached and made into specialized organizational systems, then – depending on the nature of the nation-wide economic processes which are being brought into the co-operative system – these sectors have to be built either on the principle of a single centralized organization or alternatively they have to be built as a federation, where the work of local organizations is entirely independent of the centres and where the associations serve the local organizations in accordance with the latter's requirements.

It can easily be understood that the co-ordination of the work of a centralized marketing co-operative system, and the maintenance of unity within an organization which includes scores of associations and co-operatives, is possible only on the basis of internal discipline and co-operative solidarity amongst those who belong to the organization.

The main basis for this solidarity is, of course, the sense of co-operative awareness. It has become customary for all the textbooks to say this. But we have no doubt that this awareness needs to be reinforced by economic sanctions – in the form of real economic pressure exercised by the centres and the associations on their membership.

This discipline can be based on three factors, namely:

1. The complete ban on selling a product otherwise than through the centre.
2. The actual profitability of selling through the centre.
3. The possibility of disciplinary measures by the centre against the local associations.

The first two factors usually result from the very nature of the market and the position which the central co-operative organization occupies in the market. The third factor operates because the granting of credit for co-operative marketing operations has been entirely handed over to the centre. Given the financial weakness of local associations, this has proved to be a very impressive and effective disciplinary weapon.

Such are the emergency powers which the co-operative movement has conferred on its centres, for use if the need arises. Economic necessity would have obliged us to surrender these powers; and we do not dispute the expediency of doing so. But we must recognize that these powers – like emergency powers and states of emergency of whatever kind – are fraught with exceedingly grave dangers.

Co-operative centres that possess such powers could easily be deflected on to the path of pure entrepreneurship; and the concern with enterprise could stifle co-operation as a social movement and thus undermine its sources of inspiration. Therefore, when we endow our centres with emergency powers we must take adequate steps to ensure that no one within our ranks ever forgets that co-operation is not merely a co-operative enterprise: it is also co-operative movement.

Unfortunately, we, the practitioners of the co-operative movement, preoccupied with building up our competitive strength in the struggle against international commercial capital, and engaged in the work of perfecting our central organization, have devoted too little attention to the co-operative grass-roots, where the social forces evolve which provide the driving force for our work. And the need to remember our character as a mass movement is, particularly at the present time, utterly imperative.

During the critical moments of our revolution, and also of the great French revolution, when the state was in danger and the state apparatus came under the blows of its enemies, the people's leaders more than once proclaimed the slogan 'To the masses!', and they hurled into the struggle the spontaneous forces of the popular movement which saved the situation.

As we work to perfect our entrepreneurial apparatuses, we must clearly remember that critical periods may arise in the development of our economic life, when the only salvation will lie in the conscious, or perhaps the spontaneous, capacity for resistance among the masses involved in the co-operative system.

At that moment, when all entrepreneurial methods prove powerless, when economic crisis as well as the blows of the adversary organized from abroad, wipe out our elaborate organizations, there is only one reliable path of salvation open to us – a path which is unknown to capitalist organizations and is beyond their reach. This is the path which involves deflecting the weight of the blow on to those submerged foundations upon which the whole of our work depends: on to the peasant economy, with its tens of milliards of roubles of capital, its labour force, its capacity for resistance and its awareness.

And in order that the peasants should not shun this burden, they have to feel, know and become accustomed to the fact that the cause of agricultural co-operation is their own cause, the cause of the peasants! And this cause must itself become a genuinely powerful social movement – and not a mere enterprise!

There has to be a co-operative peasant public opinion in the countryside and a mass involvement of the peasant masses in our work. Otherwise, co-operation will always be in danger and in a state of unstable equilibrium.

The thoroughgoing involvement of the masses in the co-operative system of the vertical concentration of agriculture is all the more important for us, who are striving to introduce an element of planning into the structure of the national economy, because this involvement is the only effective means of linking the spontaneous activity of the many millions of peasants with the structure of a planned state economy. This involvement of the peasant masses in co-operation is the only method which can, through prolonged work, turn our diffuse individualistic agriculture into a powerful economic system which, when combined with state industry, is alone capable of becoming the starting point for the building of the economic foundations of a future socialist society.

Petrovsko-Razumovskoye,
14 November 1926